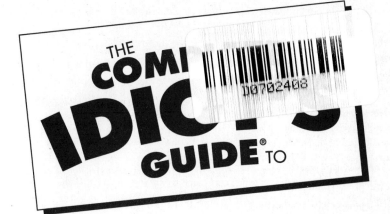

THE COMPLETE IDIOT'S GUIDE® TO

Disaster Preparedness

*by Dr. Maurice A. Ramirez, DO,
and John Hedtke*

ALPHA

A member of Penguin Group (USA) Inc.

To Chief Elaine S. Hedtke and Lt. Stephen L. Mauer,
my two favorite police officers. —John

ALPHA BOOKS

Published by the Penguin Group

Penguin Group (USA) Inc., 375 Hudson Street, New York, New York 10014, USA

Penguin Group (Canada), 90 Eglinton Avenue East, Suite 700, Toronto, Ontario M4P 2Y3, Canada (a division of Pearson Penguin Canada Inc.)

Penguin Books Ltd., 80 Strand, London WC2R 0RL, England

Penguin Ireland, 25 St. Stephen's Green, Dublin 2, Ireland (a division of Penguin Books Ltd.)

Penguin Group (Australia), 250 Camberwell Road, Camberwell, Victoria 3124, Australia (a division of Pearson Australia Group Pty. Ltd.)

Penguin Books India Pvt. Ltd., 11 Community Centre, Panchsheel Park, New Delhi—110 017, India

Penguin Group (NZ), 67 Apollo Drive, Rosedale, North Shore, Auckland 1311, New Zealand (a division of Pearson New Zealand Ltd.)

Penguin Books (South Africa) (Pty.) Ltd., 24 Sturdee Avenue, Rosebank, Johannesburg 2196, South Africa

Penguin Books Ltd., Registered Offices: 80 Strand, London WC2R 0RL, England

International Standard Book Number: 978-1-59257-893-1
Library of Congress Catalog Card Number: 2009920701

11 10 09   8 7 6 5 4 3 2 1

Interpretation of the printing code: The rightmost number of the first series of numbers is the year of the book's printing; the rightmost number of the second series of numbers is the number of the book's printing. For example, a printing code of 09-1 shows that the first printing occurred in 2009.

Printed in the United States of America

Note: This publication contains the opinions and ideas of its authors. It is intended to provide helpful and informative material on the subject matter covered. It is sold with the understanding that the authors and publisher are not engaged in rendering professional services in the book. If the reader requires personal assistance or advice, a competent professional should be consulted.

The authors and publisher specifically disclaim any responsibility for any liability, loss, or risk, personal or otherwise, which is incurred as a consequence, directly or indirectly, of the use and application of any of the contents of this book.

Most Alpha books are available at special quantity discounts for bulk purchases for sales promotions, premiums, fund-raising, or educational use. Special books, or book excerpts, can also be created to fit specific needs.

For details, write: Special Markets, Alpha Books, 375 Hudson Street, New York, NY 10014.

Publisher: *Marie Butler-Knight*
Editorial Director: *Mike Sanders*
Senior Managing Editor: *Billy Fields*
Senior Acquisitions Editor: *Paul Dinas*
Senior Development Editor: *Phil Kitchel*
Production Editor: *Kayla Dugger*

Copy Editor: *Tricia Liebig*
Cover Designer: *Bill Thomas*
Book Designer: *Trina Wurst*
Indexer: *Brad Herriman*
Layout: *Brian Massey*
Proofreader: *John Etchison*

Contents at a Glance

Contents

Introduction

Disaster can strike anywhere at any time. In this book, we describe the steps you need to take to be prepared for the inevitable. This is more than just an emergency preparedness or survival guide. You'll come away from this book with concrete strategies and ideas for how to prepare for and endure disasters.

A disaster is any set of circumstances where your needs exceed your resources. There are many sizes and types of disasters, such as …

- **Personal:** Your house has burned down or you've suffered major financial losses.

- **Local:** A chemical spill in your community or neighborhood.

- **Regional:** Tornados, rains, flooding, wildfires.

- **State or interstate:** Rains and flooding, a hurricane, an earthquake, or a volcano.

- **National or global:** A tsunami or a pandemic such as influenza.

The key to surviving all these disasters is preparedness. Each of these disasters can be personally devastating, but they're all really the same in how they affect you and your family. Each one requires you to make substantial changes to the way you lead your life. What you do for any one of these may be different, but they all require similar strategies: you have to make a plan for how to deal with disasters.

Effective disaster planning isn't trying to think of all possible disaster scenarios. Instead, you have to identify your needs and figure out how to meet them. It's learning how to develop a sense of resilience. Confidence is the key to your ability to survive disasters of all kinds. You can develop your confidence by knowing what to do and when to do it. Being able to make a decision and act on it quickly will often carry the day.

Disasters are almost always unpredictable and unexpected, but it's a virtual certainty that you will have to deal with a disaster in your life. When you understand how easy it is for disaster to strike, you can start creating plans and developing a resilient attitude. By showing you how to prepare for the unexpected and be ready for the world turning upside down, we can increase your chances for survival and even your continued prosperity in circumstances that might otherwise have seemed impossible.

This book is divided into three parts:

In **Part 1, "Disaster Preparedness: The Basics,"** you'll learn what disasters really are. Disasters aren't individual happenings; they're all the same thing. What you need to do is to identify your particular needs and then develop a plan that supports those requirements. As part of this, you have to create an emergency kit that you can grab on the way out the door. You also have to make provisions for special needs: children, seniors, people with disabilities, and pets. This part also teaches you about the most important aspect of surviving during a disaster, taking shelter. You also need to plan for business survival and continuity so you have a job and a sense of normalcy when everything's over.

Part 2, "Detailing the Disasters," provides you with detailed information about a number of common disasters that you may face at one time or another: floods, hurricanes, tornadoes, wildfires, earthquakes, blizzards, pandemics, nuclear meltdowns, toxic spills, and bombings. Each disaster brings special kinds of hazards to be dealt with. Although you probably won't meet more than a few of these during your life, you need to be ready for the possibility of any or all of these. You'll learn how to survive each of these disasters, including how to prepare for the associated dangers and problems.

Part 3, "Post Disaster," teaches you the final part of disaster preparedness: getting back to your life. After you've survived the immediate disaster, you need to restore the patterns of your life that make it your life. This includes returning to your home if you evacuated, cleaning up damage and dealing with insurance companies, and going back to work. There are things you can do to deal with disasters even when you made no plans at all. (Knowing that you have options is a big help.) You'll also be able to revise your plans based on how well your current plan worked, so next time things will go even better.

Extras

As you read the text, you'll notice the occasional sidebar. There are four types:

Life Preservers

Tips for making effective preparations before a disaster strikes.

def•i•ni•tion

Short, snappy explanations of disaster and disaster preparation terminology.

Storm Warnings

Warnings about potential problems or situations to avoid.

Prep Facts

Miscellaneous facts about disasters and disaster planning.

Acknowledgments

This book is the culmination of my belief that the American people and all people are the best caretakers of their own destiny and security.

I must first thank and acknowledge my co-author, John Hedtke, whose expertise and craftsmanship took ideas and made them a book.

I also thank my wife, Laura, my daughters Victoria and Tiffany, and my sons Nicholas and Christopher, who make it possible for me to serve the people of this great nation and learn the lessons brought to you in this book.

Special thank yous go to:

- My friend and teammate Allison Sakara, MSN, ARNP, who taught me French and who compiled the information in Appendix A, "Additional Resources."

- My friend and shipmate Cmdr. Pietro (Pete) Marghella, MS, MA, CEM, FACCP (USN Ret.), who taught me everything I know about strategic thought and planning.

- My friend and colleague James Shultz, Ph.D., who mentored me to expertise in Disaster Behavioral Health.

I also must acknowledge those with whom I serve when the nation calls, the men and women of DMAT-FL3 and the National Disaster Medical System and those with whom I served before. No one forgotten, no one left behind …

—Maurice

This book has been a real pleasure. I'd like to first acknowledge my co-author, Maurice Ramirez, without whom this book wouldn't have happened in the first place.

I'd also like to thank my wife, Marilyn, and my youngest stepdaughter, Susan the Wonderchild, who put up with (in reasonably good grace) my hiding in the office for days and nights at a time while churning out deathless prose.

And to our many cats—Yang, Yin, Bo, Silas, Sebastian, and Bernie— who erased a week of e-mail by walking on the mail computer's keyboard, who randomly attacked any piece of paper left on the floor or my desk, and who climbed my legs with their needlelike claws just so they could be closer to me: thanks, thanks a yahoo.

—John

Trademarks

Part 1

Disaster Preparedness: The Basics

Everyone thinks about disasters as unique and separate occurrences. You'll be surprised to find out that they're all really the same. Sure, you do some different things for each type of disaster, but the basics are always the same. In this part, you'll learn what a disaster is, how to build a plan that meets your requirements, and how to be ready at a moment's notice. You'll also see how to take shelter (both at home and in a public shelter) and what you may need to do for children, seniors, pets, or people with disabilities. In addition, there's information about how to prepare your business so it's able to continue as soon as the dust settles.

When Needs Exceed Resources

In This Chapter

- ◆ Preparing for disaster
- ◆ The difference between a disaster and a catastrophe
- ◆ All disasters are the same
- ◆ Recent disasters
- ◆ Calculating the physical and emotional risks of a disaster
- ◆ All disasters happen locally

To plan for disasters, you need to understand what you're planning for. There are many different types of disasters, some natural, some accidental, and maybe even some intentional. All of them disrupt the processes of your life, put you and your family and friends at risk, and can cause serious financial losses. Knowing that you can face danger at any time is the first step toward being ready to overcome and survive it.

In this chapter, I tell you about public awareness of disasters and disaster preparedness; the parameters of disasters including examples of recent disasters; how to calculate the risks and the damage from disasters; and that all disasters are local, no matter how far-ranging they may be.

Disaster Preparedness: The New CPR

When cardiopulmonary resuscitation (CPR) was invented in the 1970s, the goal was to train as many potential bystanders as possible to help if someone had a heart attack or choked in public. In an effort to educate everyone about the importance of learning basic chest compression and the Heimlich maneuver, even Hollywood got in on the act, incorporating the practices into movie and TV storylines. As a result of great marketing, virtually everyone knows what CPR is, and hundreds of thousands of people are trained to do it.

In the new millennium, a heightened awareness of both terrorism and the impact of natural disasters has created a need for a new CPR: core skills that will help us meet the challenges of man-made and natural disasters. Why is this important? Consider the following:

◆ The 1994 Northridge, California, earthquake wiped out 8 hospitals and affected 20 million people.

◆ In 2008, Hurricane Ike decimated Galveston Island and much of the Texas Gulf Coast.

◆ In 2005, Hurricanes Katrina, Rita, and Wilma decimated much of three major Gulf Coast cities.

◆ In 2004, Hurricanes Charley, Frances, Ivan, and Jeanne laid waste to Florida.

◆ No one will ever forget the World Trade Center bombings on September 11, 2001.

Ironically, many people believe they need CPR training more than they need disaster preparedness. In fact, you are far more likely to face a disaster at some point in your life than you are to be a bystander when someone experiences a heart attack.

Like heart attacks, disasters have always happened. We're more aware of disasters and are called upon to respond as never before. The number of people in the last decade who have been directly affected by natural disasters is double the number of people who have experienced heart attacks. So it's twice as likely that you, your family, and your neighbors will need disaster preparedness skills as opposed to ever needing your CPR skills.

Parameters of Disaster

The parameters of disaster are the size, the scope, and the extent of the disaster. The size of a disaster is just that: how big an area the disaster affects. Disasters can be ...

- Local—affecting just a few blocks or a community, such as a creek overflowing and flooding a small area.

- Municipal—affecting a city or county, such as a toxic spill from a train derailment.

- Regional—affecting several counties or part of a state, such as wildfires in Southern California.

- Interstate—such as Hurricane Katrina or Hurricane Ike.

- National—such as a flu pandemic.

The scope of the disaster is the effect of the disaster and the number of people being affected. Is the disaster affecting just people, people and property, or property only? Is the disaster causing physical injuries or is it also causing emotional damage? A tornado may cause significant damage to property throughout a community or even a city, but the number of people being injured or emotionally damaged may be relatively small. On the other hand, a wildfire might burn whole blocks or even portions of a city, causing substantial disruption to established communities, making the scope much larger.

Finally, the extent of a disaster is the depth of the disaster's effects. The extent of the damage to an area hit by a flood is fairly severe. However, a toxic spill of chlorine gas from a train derailment that didn't cause serious injuries and had no lingering effects after the gas dissipated would have a much lesser extent.

Definition of Disaster

In the simplest of terms, a disaster is any situation in which needs exceed resources. People usually think of a disaster as a specific type of *incident*: a flood, a tornado, an earthquake, etc. These are certainly disasters, but they're all situations in which the resources you can muster in response aren't sufficient to the needs and demands of the situation. In a disaster, an incident occurs or a situation develops where you know what the necessary response is, but the sheer magnitude of the need outstrips the people, supplies, and time available to mount that response.

def•i•ni•tion

When you're talking about disasters and disaster planning, **incidents** are accidental or unplanned occurrences.

Preparing for disaster means that you are reversing the disaster equation: you are increasing and conserving your resources against the day when you have a great need. This mitigates disaster by ensuring that your resources will exceed your needs.

Definition of Catastrophe

A catastrophe differs from a disaster in scale. A catastrophe is any situation in which needs exceed *all* ability to respond. A catastrophe can occur in two ways. The first is a situation where you know what the response is, but the need is too great for any resource preparation to mitigate. For example, an 8.1 earthquake in Southern California or a cluster of tornadoes hitting downtown Houston would be catastrophic. No amount of planning or preparation would make any significant difference in the devastation that would result from these situations.

The second kind of catastrophe is a disaster that gets out of hand. This is a situation where you don't know what the appropriate response is or you haven't adequately practiced your response. In this case, even though the magnitude of the need is less than the available resources, you've lost the ability to respond effectively. You might have an opportunity to cut off a wildfire from spreading into a nearby housing development by cleaning out the brush in between the homes and the fire by simply fielding enough firefighters in the area to shovel dirt on the fire and smother it. But if you didn't know how to clean out the brush or your teams were inexperienced and didn't make the firebreak wide enough, the fire could spread through a whole new area, with no way of extinguishing it: a catastrophe.

Reversing the disaster equation is only part of preparing for a catastrophe. Preparations also require planning for the unexpected and practicing the basic skills of disaster response. Fortunately, although the specific hazards you may face differ, all disasters are really the same. They magnify needs in common ways and along common paths, regardless of the specific cause of the disaster. As a result, the bulk of planning and preparation for disasters of all kinds is the same. You don't have to make one set of plans for a tornado, a second set for wildfires, and a third set for toxic spills. Effective disaster planning tends to be a comprehensive, "one-size-fits-all" kind of thing.

Great Disasters in Recent History

Regardless of where you live, you're at risk for disasters. Live in the mountains? Blizzards, avalanches, quakes, and even the occasional volcano could happen. Live on the plains? Blizzards, wildfires, floods, and tornadoes are possible. Live near the coast? You're at risk for hurricanes, storms, and tidal waves. And no matter where you live in the United States, there is a chance of toxic spills and nuclear meltdowns.

Disasters strike anywhere, anytime. Some of the notable recent disasters have included fires, floods, earthquakes, hurricanes, and tidal waves. It doesn't matter where you live—there's a chance that a disaster of some kind can occur.

The Southern California Wildfires

Fire is part of the natural cycle for a number of native plants in Southern California. Almost every year, small wildfires may burn several thousand acres and possibly a few houses, but active firefighting is usually adequate to keep things under control. However, in October 2007, a series of wildfires consumed half a million acres throughout Southern California.

The wildfires had multiple origins, including suspected arson, downed power lines, a construction accident, a structure fire, and a truck accident. Extreme drought, high temperatures, and strong Santa Ana winds gave the fires strength and spread the flames quickly from area to area. The fires were finally contained almost three weeks after they began, thanks to the efforts of more than 10,000 firefighters, National Guard units, U.S. Armed Forces units, prisoners, and volunteers. More than 1,500 homes burned and 9 people died. The Southern California wildfires of 2007 were also responsible for the largest evacuation in the United States, about a million people.

Other notable fires in recent years include:

♦ The 2007 Zaca fire in Southern California

♦ The 2007 Cedar fire in Southern California

♦ The 2007 Milford Flat fire in Utah

♦ The 2004 Alaska fires

♦ The 2003 Aspen fire near Tucson, Arizona

♦ The 2002 Biscuit Creek fire in Southern Oregon and California

Wildfires don't occur only in the western states. Any state where there are extended hot, dry periods can and will have wildfires.

The Hundred-Year Flood

There have been a number of extreme floods in the Midwest, most of which are related to the Mississippi and the Missouri river basins. These are inevitably dubbed *hundred-year floods* in the news because of their severity. A century doesn't last as long as it used to.

def•i•ni•tion

A **hundred-year flood** is the level of flood water that happens—on average—every 100 years. You can have multiple hundred-year floods in a century or even in a given decade.

The floods in June 2008 in the Midwest were exceptionally bad. These were brought on by heavy rains for several months, causing rivers to overflow their banks in a number of areas in the Upper Midwest.

The worst of the flooding happened in Iowa, but parts of Illinois and Missouri were also hard hit. Thirteen people died as a result of the floods. This was the worst flooding to hit the area in 15 years, when heavy winter snowfall and heavy spring rains drenched the upper Midwest with dramatically higher amounts of rain than normal.

Other recent floods that have caused significant amounts of damage are:

- The 2007 Northwest Coast floods in Washington and Oregon
- The 1997 Red River floods in North Dakota
- The 1996 Willamette Valley flood in Oregon
- The 1995 Central California floods

The Six-Minute Earthquake

The Pacific Rim area has always been prone to earthquakes, with a significant quake hitting somewhere almost every year. The 2008 Chengdu earthquake in Sichuan Province, China, struck three months before the 2008 Chinese Summer Olympics. At least 69,000 people were reported killed and close to 5 million people were left homeless.

An even larger quake in December 2004 off the coast of Sumatra was responsible for the tsunami that killed at least 250,000 people and damaged buildings and coastlines in a number of Pacific Rim countries.

Other recent earthquakes worth noting include:

- The 2008 Illinois earthquake
- The 2006 Hawaii earthquake
- The 2002 Denali earthquake in Alaska
- The 2001 Nisqually earthquake in Washington
- The 1994 Northridge earthquake in Southern California
- The 1989 Loma Prieta earthquake in the California Bay Area

Fourteen Months of Hurricane Horror

Katrina is one of the best known as well as the costliest hurricane to have ever hit the United States, but it was not the only—nor even the strongest—in a 14-month string of hurricanes to hit the Gulf Coast.

The first hurricane in the string was Charley in August 2004, which was at the time the strongest hurricane to hit the U.S. coast since Hurricane Andrew in 1992. It caused about $15 billion in damage. A few weeks later, Hurricane Frances hit Florida. Although Frances was a much slower hurricane, it lingered and caused another $8 billion in damage to the area. Two weeks after that, Hurricane Ivan did similar damage in Alabama. A week and a half later, Hurricane Jeanne hit the same part of Florida that had been hit by Frances; fortunately, it "only" caused about $3.5 billion in damage. The 2004 hurricane season finally ended, but 2005 would not be kind either.

In August 2005, Hurricane Katrina clobbered the Louisiana coast, flooding New Orleans, killing more than 1,800 people, and causing evacuations and $81 billion in storm damage along most of the Gulf Coast. A month later, Hurricane Rita, which was even stronger than Katrina, hit the Texas and Louisiana coastlines. Rita caused the evacuation of 3 million people in Texas (the largest evacuation in Texas history) and caused $10 billion in damage. The final major hurricane of the 2005 season was Hurricane Wilma, the largest hurricane in Atlantic history. Wilma caused $20 billion in damages.

By the end of this string of hurricanes, many areas were badly damaged and the citizens were emotionally drained. People can and do recover from a single hurricane, but the continual pounding across 14 months sapped many people's emotional resilience and severely depleted the disaster response resources for communities and states.

The 2005 hurricane season was the worst on record, but warming Gulf Coast waters are likely to produce similar seasons in the future. But even with unchanged conditions, severe hurricanes hit the area regularly:

◆ Hurricanes Dean and Felix in 2007

◆ Hurricane Isabel in 2003

- Hurricane Mitch in 1998
- Hurricane Andrew in 1992

Disaster Vulnerability

Your vulnerability to disasters depends on a lot of different factors. As the preceding lists show, every community has the potential for disasters. But how likely you are to be hit by a disaster is something else. If you live near the coast in Southern California, there are real dangers of earthquakes, wildfires, floods, and tsunamis, but not blizzards. On the other hand, if you live in the mountains in North Dakota, you might have wildfires occasionally and you'll definitely have blizzards, but never tsunamis.

The "Footprint" of Disaster

The word "disaster" first brings to mind physical destruction: images of shattered homes, burnt communities, flooded streets, wind-splintered trees, and toppled towns. The footprint of disaster includes not only the geographic area encompassing the area of physical destruction, but the much larger region that is damaged emotionally by the loss suffered as a result of the disaster. The actual region of physical destruction at Ground Zero in New York City is notable for its limited size, yet an entire nation and the entire world was impacted emotionally.

Risk and Outrage

The footprint of disaster defines the two aspects of disaster vulnerability: the risk of physical destruction and the outrage over disaster-related suffering.

Risk is the likelihood that a disaster will occur. It is the product of the statistical probability that some disaster will occur and the hazard score for the disaster in question. You can calculate the hazard score for any given disaster or derive it from the actuarial data used to rate property insurance in a given area. To calculate the hazard score, you must also determine the impact a disaster would have if it occurred and your vulnerability to that particular disaster.

Outrage is the emotional impact of a disaster. For example, no matter where you are, there is a certain amount of risk for flooding or wildfire. (Your insurance company can give you an idea of what the risks for these hazards are.) You may be at relatively low risk for both types of disaster, but the emotional impact of having your community flooded or burned could be extremely high if you've lived there for several decades or extremely low if you were living in an apartment temporarily and hadn't actually moved into the area yet. It all depends.

Calculating Your Risks

You can actually get a general idea of how susceptible you are to a disaster and what the public sentiment will be using the following calculations. Because you're dealing with the estimated emotional impact of a given disaster, your calculations are going to be very subjective, but you can still use these numbers to get a sense of what might happen.

Start by identifying the *probability* of a disaster. The probability is the likelihood of a disaster occurring, from 0 to 100 percent. For this kind of information, you can talk to your insurance company, check with the local library, look online, or even talk to the local chamber of commerce. Check past disasters and learn what types that area may be subject to.

For example, the probability for a hurricane or tropical storm hitting Florida is 100 percent for any given year. The chances of a really powerful, destructive hurricane hitting Florida in any given year are much less, perhaps only about 30 percent. Similarly, the chances for a powerful earthquake hitting somewhere on the Pacific Rim in a given year are 80 percent. The chances of a powerful earthquake hitting near enough to where you live on the Pacific Rim to do significant damage are much lower, probably substantially less than 1 percent in any given year.

The next factor to identify is the *impact* of an experience on a scale of 0 to 3 (where 0 = no impact, 1 = minimal impact, 2 = moderate impact, and 3 = significant impact). For example, the impact of a wildfire that swept through areas 15 to 20 miles away might be nothing at all, but it could be moderate if you were forced to evacuate or even significant if you lost houses in your community or your own house.

After this, you need to calculate your *vulnerability* to a disaster on a scale of 0 to 3 (where 0 = not vulnerable at all, 1 = minimally vulnerable, 2 = moderately vulnerable, and 3 = extremely vulnerable). If you're single and live in an apartment with no pets, you may not be vulnerable to disasters at all, but if you have large animals or family members who require special medical care, you may be extremely vulnerable to disaster.

The next thing you need to calculate is the potential for *outrage:* how you perceive the experience on a scale of -3 to 3 (where -3 = very pleased, -2 = moderately pleased, -1 = mildly pleased, 0 = neutral, 1 = mildly outraged, 2 = moderately outraged, and 3 = very outraged).

def•i•ni•tion

Although there are many other emotions that make up the emotional impact of a disaster, **outrage**—anger, frustration, and grief—is the primary emotional impact.

Identify your *expectation*, what you think a situation should be, and your *experience*, what the reality of the situation is, on a scale of 0 to 3 (where 0 = no difficulties, 1 = minimal difficulties, 2 = moderate difficulties, and 3 = extreme difficulties). Your outrage in a given situation can be calculated as the expectation minus the experience.

For example, you may expect that evacuating and staying in a public shelter will be extremely difficult, but your actual experience may show that there were minimal difficulties. Your level of outrage is calculated as this:

Outrage (2) = Expectation (3) – Experience (1)

Calculating your potential outrage provides an interesting insight. Your expectation is the perception of what the reality should be while your experience is how things are. Outrage is simply the difference between expectations and reality.

The next calculation is the *hazard*, which can be a number from 0 to 6 (where 0 = no hazard and 6 = extremely hazardous). The hazard measures your total sense of exposure to a disaster and is calculated as this:

Hazard = Impact + Vulnerability

For example, if a wildfire is unlikely to become a direct and immediate danger to you and you're not worried about it, the hazard will be very low. On the other hand, if you're very worried about earthquakes even though the chances of earthquake damage are only moderate in your area, the hazard is fairly high.

The *risk* is the likelihood of taking any kind of damage—to your possessions, person, or psyche—from a disaster. Risk can be a number from 0 to 6 (where 0 = no risk at all and 6 = extremely risky).

Risk = Probability × Hazard

For example, if you have a 30 percent probability of a major hurricane hitting your area in a given year and the hazard is 4, your overall risk is 1.2 out of 6, which is not a lot of risk in any given year.

Finally, you need to calculate the *tolerance*, which is the measure of the emotional response to a disaster.

Tolerance = $\text{Risk}^{\text{Outrage}}$

When outrage is 0—that is, when your expectations match your experience—you're not surprised by what happened, so there's no real sense of outrage. When outrage is 0, the tolerance is always 1. (Mathematically, any number raised to the power of 0 equals 1.) When your outrage is a positive number because your expectations were greater than your experiences, the tolerance score is a reflection of your anger and frustration. However, when your outrage is a negative number because your experiences exceeded your expectations, the tolerance score is a reflection of your enthusiasm and satisfaction.

Calculating the tolerance can also give tremendous insight into why a large-scale disaster can result in little national concern, while a physically limited disaster can become a full-fledged national catastrophe. The public's expectations for how a disaster will be dealt with can be confounded by the reported experiences and there can be vicarious outrage. This was a common reaction after Hurricane Katrina, when the expectations were that the state and federal government's efforts would be much more effective and timely than they appeared to be, resulting in a national sense of outrage at the government's response.

All Disasters Are Local

As part of your personal preparations at home and work, it is important to know what assistance you can expect from local, state, and federal officials as well as from nongovernmental organizations such as the American Red Cross. It takes time to mobilize governmental and nongovernmental assistance, and even more time to transport people and resources to your disaster. This creates a limitation that you must accommodate in your disaster plans.

Acknowledging that all disasters are local means that different localities have both different probabilities of a specific hazard occurring and different expectations for the response to the disaster. These differences alter the calculations for vulnerability. Simultaneously, treating all disasters as local reminds disaster planners (and you) that the first 72 hours of any disaster response will be in the hands of local officials, local organizations, and (most importantly) you.

Neighbors Helping Neighbors

Modern disaster preparedness is based on the idea of concentric rings of response and resources. Those resources closest to the site of the disaster are pressed into service first, while all resources in the successive rings begin to mobilize toward the epicenter of the disaster. Ideally, local resources are sufficient to meet needs until additional resources become available. Unfortunately, many communities are ill-prepared to support their local citizenry for the 24 to 48 hours it takes for state assets to arrive and be distributed. In many cases, ad hoc donations by water and food suppliers are the sole source of these most vital of supplies. This is the reason that personal preparedness and the personal Go-Pak are so vital for every member of your family.

Ideally, local response concentrates on maintaining and restoring municipal services (utilities, police, and firefighters) and community services (emergency rooms, hospitals, health departments, grocery stores, and pharmacies). Despite providing much of the space in public buildings for disaster response services, most municipalities actually rely on other organizations and state and federal agencies to provide shelter services, food distribution, water distribution, and disaster health care.

Hand, Heart, and Cross

The modern era of volunteer groups, nongovernmental organizations (NGOs), and faith-based organizations provides a community with an array of disaster-response services. In many cases, these disaster responders have decades of personal and organizational experience. Some organizations, such as the American Red Cross, deploy resources and personnel ahead of predictable disasters, staffing shelters, providing water and meals, and even caring for the emotional impact of the disaster. Other organizations, including the Southern Baptist Chain Saw Caravan, respond after the event with teams trained and experienced at the specific task of clearing fallen debris. Insurance companies have disaster teams charged not only with claims adjustment, but assisting individuals with other, noninsurance needs such as referral to medical care, housing, and even water distribution.

Although these resources are seldom housed locally in the community, the independent command structure of the NGO and similar groups allows them to move faster than more hierarchical organizations or agencies. When in place in the disaster-response area, these organizations reach out to local, state, and federal response agencies to coordinate and supplement the governmental initiatives.

I'm from the Government

Government response is based on a series of policies including the National Response Framework (NRF) and the National Incident Management System (NIMS). All federal agencies are required to participate in and conform to the structure of the NRF and NIMS. State governments receiving federal disaster-preparedness and response funding are required to conform to a similar structure as a condition of that funding. The NRF and NIMS are not a formal response plan, but rather an "all-hazards" framework for responding to disasters of both a predictable and unpredictable nature. The NRF and NIMS include 18 national-planning scenarios that encompass the disaster-response skills and historically proven needs.

The government response is divided into five broad areas that span the life cycle of disaster from planning, preparation, response, and recovery. These areas are: incident command, operations, logistics (resource management), planning (including adapting the plan during a response), and finance.

These broad areas are applied in two distinct and separate arenas: property and infrastructure, which is headed by FEMA (Federal Emergency Management Agency); and medical and humanitarian, which is headed by NDMS (the National Disaster Medical System). Many states have similar or corresponding state-based agencies to provide additional assistance to survivors of disasters. Whether state or federal, however, these agencies must assess needs and coordinate with local governments before they can start work.

Even with a green light at every level, it takes a while to get everything moving. The mobilization process typically requires a minimum of 72 hours after a disaster to be fully operational. In a large disaster, it may require an additional 48 hours to bring sufficient resources to bear to provide significant relief.

Disaster response at all levels concentrates on supplementing the restoration of municipal services and providing temporary community services, especially medical care. State and federal agencies also provide much-needed financial assistance to speed the restoration of home and community. In addition, state and federal disaster response coordinates among the various governmental and nongovernmental response organizations through a centralized command system.

The Least You Need to Know

- Disaster is when needs exceed resources.
- Catastrophe is when needs exceed all ability to respond.
- The footprint of disaster is larger than just the physical impact and includes the psychological impact as well.
- All disasters are local.
- Life is risky. It's best to be ready for things that can happen.

2

One Size Fits All Disasters

In This Chapter

- ◆ What is disaster-specific planning?
- ◆ What is needs-based planning?
- ◆ How the two types of plans stack up against each other
- ◆ The goals of a needs-based plan
- ◆ New technologies that can make it easier to survive disasters

Disaster planning isn't about identifying the types of disasters that may befall you. Effective disaster planning identifies the services that must continue without interruption and then shows you how to keep them going. This may be a new way for you to think about disaster planning, but it's the best way to create flexible, adaptable plans that you can use regardless of the type of disaster.

In this chapter, I discuss disaster-specific versus needs-based planning, the things you need to consider when creating a needs-based plan, and ways to use (and sometimes avoid!) technology.

Keeping It Simple

Although a disaster plan will never be a single page, you should keep it as simple and straightforward as possible. The information in the plan must be easy to find and easy to read. Depending on how the plan is going to be used, you may also want to have the plan formatted so that you can hand out separate sections to people in charge of various tasks.

Preserving the Processes of Life

All disasters disrupt the processes of your life—the everyday routines of your household, your family, your business, and your community. The underlying objectives of all disaster planning are to preserve the processes of life and to continue living as close to your usual life as you can. Since you can't predict what's going to happen, think of the basic life functions you want to sustain. Your choices of how to support these functions may vary from environment to environment.

Life Preservers

No matter how big your list of specific disasters, you can't plan for everything. You'd have to include asteroids, global warming, fire ants, and spontaneous human combustion. Plans that only address specific disasters are unwieldy. Worse, when something unexpected happens, you have no plan.

Disaster-Specific Plans

Disaster-specific plans are sets of steps created for one type of disaster, such as a hurricane plan or a wildfire plan. Disaster-specific plans are reactive: they focus on external occurrences or conditions to trigger an action on your part. While it may sound sensible to customize your plan and your actions to the situation, experience has shown that having separate plans for each possible type of disaster can backfire. This approach produces a large, complex encyclopedia of separate plans that rapidly becomes unwieldy and unusable.

Although it's important to include some elements of a particular disaster in your preparedness planning, preparing for and responding to one type of disaster will be different from one person to the next. Any cookie-cutter disaster plan will have inherent shortcomings such as the following:

- They don't address individual situations. For example, there would be a single approach for sheltering in place whether you live in a high-rise apartment building or a mobile home.

- They don't allow for personal priorities, such as a person's desire first to save the life of a family member or pet before preserving their house.

- They aren't adaptable to more complex situations, such as having multiple hurricanes ravaging the same area in a short time frame or if two different disasters occur at the same time (such as an earthquake followed by a flood).

Needs-Based Planning

Needs-based planning is proactive disaster planning. It focuses on how to preserve and support the processes of your life rather than on addressing specific disasters or how to mitigate something once it's already begun. Needs-based planning answers the question, "What do I need?" by identifying the goals and resources, rather than specific disaster scenarios, and by learning a single set of skills and concepts that you can generalize to all disaster situations.

Because you're working to maintain your life and get back to normal as quickly as possible, it doesn't matter what the disaster is. This makes needs-based planning substantially more effective than disaster-based planning at dealing with any kind of adversity.

As you develop a needs-based plan, you can still incorporate disaster-specific information, but you want to focus on generalized

def•i•ni•tion

Needs-based planning is disaster planning that focuses on how to preserve and support the processes of your life rather than on addressing specific disasters. Disaster-specific plans are a set of steps for a specific type of disaster, such as a hurricane or fire.

skills and concepts. This prevents you from worrying about "How do I get out of the house quickly during a tornado?" when you just have to use the same information you mentally filed under "escaping a fire." The most important steps in creating a needs-based plan are:

- Identify the goals for your success and the processes that are most important for you to support.

- Identify the resources you have available (and knowing that you'll never have enough resources, how you can best apply them).

- Practice your plan so it becomes familiar and more automatic for you and others in your household, decreasing the overall stress of the situation.

- Brainstorm possible failures of your plan and how you could adapt each failure with an alternate course of action.

- Be flexible and expect the unexpected, so you won't be over-whelmed if the situation changes course or if your plans don't work out.

Goals of Needs-Based Plans

If you've prepared a needs-based plan and practiced it, your needs-based plan will give you the tools you need to maintain your processes without significant interruption. In fact, the very best disaster plan will let you operate as if there is no crisis whatsoever. A needs-based plan must help you do the following five tasks: protect, provide, unite, inform, and act.

Protect

The first thing a plan needs to do is protect people: you and your tribe, the people you're closest to. Protection is just that: ensuring that you and your tribe are safe. Your tribe can include family, friends, even pets. They must be safeguarded against the elements and outside harm. Knowing your tribe is safe represents the difference between being uncomfortable at being physically separated from them and being comfortable. At your core, when you know the tribe is safe, you're willing to venture away from them.

If this sounds like something from the Stone Age, you're right. It's exactly how people have dealt with protecting the ones closest to them forever. You won't leave your tribe until you're sure that everything will be okay when you return. Until that time, you and the tribe will do things together. This is seen in shelters: families will literally move as a group into shelters together because at a basic level, they don't feel safe enough to be separated. Fortunately, protection is now more sophisticated than "Me have sharp stick." Today, protection is more along the lines of you and your tribe obtaining resources: shelter, food, water, and so on. When you're sure that your tribe has resources and the knowledge to safely use them, you can feel confident when you leave them that they'll be able to "protect" themselves and each other.

Your tribe can grow under unusual circumstances. During the string of intense hurricanes in 2004—Charley, Frances, and Jeanne—people found themselves in the same shelter with the same people over and over during a short period. By the time Jeanne rolled around, people were showing up at shelters asking to be put next to the people they'd been next to before.

Provide

After you and your tribe have safe shelter, you can think about sustenance: water and food. Hunter-gatherers have to go out from the cave to forage. Modern foraging may be going to work, going to the Red Cross to get more food, or arranging with FEMA for blue tarps and a check. But it isn't sitting huddled in the corner; it's going out and doing things for yourself.

Again, the whole tribe probably won't go out foraging for supplies. You'll have one or two people finding water and food while the remainder stay behind. When you're set up for shelter and protection, then water and food, you can start to do things that go beyond basic survival.

Unite

People who aren't set up with shelter, water, and food aren't going to be interested in anything outside their own and their tribe's needs until those needs are met. After those needs are met, however, they're ready to start thinking about other things beyond basic survival.

As part of your disaster planning, you need to identify how the tribe will be led. In a traditional family group, one of the parents is a hierarchical leader. The other type of leadership organization is cellular, where people do things they're best suited for. For example, Dad may be in charge of everything external to the tribe. He's the one who goes out of the shelter to work or deal with the Red Cross or FEMA. Mom or some other member of the tribe may deal with domestic things. Yet another person may mind the younger children.

In fact, what you often see is a hybrid of both types of leadership. A certain amount of hierarchical leadership will happen, but if someone can do a task better, they do. For example, the 12-year-old daughter may fill out the computer forms at the FEMA desk because she's the most adept with computers and can do this fastest.

The other part of uniting is that people in the tribe need to stay connected. If I'm the dad and I don't have any way to reach the tribe, I'm not going anywhere. Everyone needs to know that everyone else is okay. Talking to each other on the phone is a surrogate for holding hands to make sure that everyone's still okay.

If you're making a disaster plan for your family, consider who will be with you and how to prepare each person for the disaster. If you have small children, you may need to talk to them about what is happening, and reassure them that everything will be all right. You also need to identify what tasks each person will perform. For example, if you're facing a hurricane, you need to say who's going to board up the windows or who is responsible for getting the dog into the car if you evacuate.

Each person should do something that ensures the safety and security of everyone else. Even children can participate. A small task will make a child feel more useful and more in control, as a critical part of the plan, rather than a helpless bystander. Depending on your children's ages, put them in charge of getting extra batteries or filling water bottles.

Similarly, if you're making a plan for your business, consider who will participate and what role each person will fill. If you plan to close the business, you need to know who will be involved in the decision to close and how you will secure the premises. If you decide to stay open, your disaster plan is even more important because you will be responsible for the safety of your employees.

Other people in your plan include contacts outside the disaster zone. You need someone to serve as a message board for communication. Then everyone involved in your plan can call in and let the central contact know they are safe and their location. If you decide to leave, you need someone out of state with whom you can stay.

Finally, consider what outside facilities you are going to rely on. If you have unanticipated emergencies, you need to know who you are going to call and if they're going to be able to get to you. Be aware that for the first 72 hours you'll be on your own. The disaster medical teams don't come for 24 to 72 hours after a disaster. The National Guard can't come until a disaster is declared, and then it usually takes another 12 to 72 hours. FEMA doesn't come for days. All of those outside supports are late events. If you haven't set up your support systems, you are going to run out of manpower and supplies. If your entire plan is to call 911 and wait, you need to reassess your plan.

Inform

Everyone needs to stay informed about what's going on, but this doesn't mean that you should shower everyone with information they don't want or don't need. When you are passing on information to people, whether they're part of your tribe, strangers, or people in a business setting, consider the following criteria:

- ◆ Do these people need to know what I'm telling them?
- ◆ Do they want to know?
- ◆ Is this information accurate, as far as I know?
- ◆ Will this worry or upset them unnecessarily?
- ◆ Will this get in the way of them accomplishing what they need to do?

Not all information should be passed on. Some news won't help anyone sleep better or help them get their jobs done. You shouldn't ever lie when passing on information, but do exercise a measure of discretion. Information that only serves to upset people and raise their stress level isn't helping anyone. If someone asks you specific details, answer their questions, but don't feel you have to volunteer everything unfiltered.

Informing is not just passing on the latest news, it also includes communicating information and procedures. When handling communications in a business situation, you need to know how to contact people and they will need to know how to contact you. You also need to have redundancy so that procedures and information essential to the continuity of your business operations aren't lost or unavailable when you need that information. As part of your planning, make sure that you document the procedures necessary to survival, and keep copies of your procedures on and off-site.

Act

Finally, when your disaster plan says that you need to do something, do it. Frequently, this can mean that you need to evacuate, but it can also mean that you need to shelter in place, start working on restoring business systems, and almost anything else.

Twenty-First-Century Disaster Preparedness

Because needs-based planning proactively seeks to support and maintain your day-to-day processes, you can benefit strongly from technology that gives you more information about a situation. Many technologies make it easier for you to support your processes, but keep in mind that although technology facilitates results, it's not a necessity. And in some cases, technology that you're used to in normal conditions can be dangerous in a disaster.

Special Preparedness Technology

One of the most effective uses of current technology is also extremely approachable. In many disasters, some of the most fragile information is documents, records, and family photos. These are easy to damage and hard to replace. Keeping them in a safe-deposit box not only makes it difficult to use them but safe-deposit boxes can be damaged by fires, floods, and other disasters as well.

You can use an inexpensive scanner or in some cases a digital camera to create digital copies of these that you can then store on CDs. There are also scanning units that scan film negatives and 35 mm slides. You can even make multiple copies of the CDs and send a storage pack to several different locations. (For family and personal photos, you can even go one step further and share these with family and friends.)

A number of organizations are working on various kinds of self-propelled robots designed to crawl over or through rubble. The robots supply audio and video information to a central computer and can have sensors for temperature, gas, and noise. They can also carry a two-way communication link so that when they reach someone who's trapped, they can talk to the rescuers. Many of these robots are only about a foot tall and are designed to crawl through rubble, but some larger robots can climb rocks or move independently underwater.

Evac-Pack, developed by researchers at the University of California, Irvine, is a sensor pack worn by disaster responders. It gathers data from the location and sends it back to a monitoring computer in the operations center. The Evac-Pack has a backpack computer and a wearable keyboard and mouse. There's an eyeglass-mounted visual display, a mike and earpiece, and a helmet that transmits real-time images, temperature, speed, and location to the monitoring computer. In addition, sensors monitor for the build-up of harmful gasses. The monitoring computer can transmit a floor map to the visual display so the wearer knows where they are as they search the building.

One of the most valuable uses of technology may be telemedicine. Telemedicine is basically remote medical examinations. It requires fairly basic equipment at the patient's site: a computer, a webcam and microphone, an electronic stethoscope, an electronic blood pressure cuff, and an Internet connection. The physician could examine the patient from anywhere else in the world. Diagrams on the screen can show the patient where to apply the head of the electronic stethoscope. With this relatively inexpensive equipment, the physician would be able to perform a basic medical examination and make a number of health-care decisions.

The advantages to telemedicine are that it provides fast, effective screening of patients so that only those needing critical care take up time in the ER. Telemedicine also minimizes unnecessary contact between patients with infectious diseases such as the flu and those who haven't been exposed to it. (This technique of avoiding contact with potentially sick people is called "social distancing" and it was very effective in limiting the spread of SARS in Toronto several years ago.)

There are a couple of problems with telemedicine. First, a telemedicine examination takes longer than an examination done in person, which can be an issue if medical resources are in short supply. Another problem is that many states have made telemedicine illegal because of a reaction to the potential for misprescribing illegal narcotics.

Despite these problems, telemedicine has been used to great effect in several disasters already, including the 2005 earthquake in Pakistan. Telemedicine is also being used to supply medical care to remote rural areas where travel to a medical facility is difficult internationally and within the United States. For example, the University of Kentucky College of Medicine offers clinical telemedicine services throughout Kentucky through their Kentucky TeleCare program.

Wireless and Internet Resources

The Internet has changed almost every aspect of our lives, and disaster preparedness is no exception.

Weather websites such as Weather.com are a powerful way to get information about upcoming weather conditions and monitor national weather alerts. There are also programs that run on your computer in the background and wake up to provide audio and visual alerts when there are weather conditions that you've specified you want to be alerted to. On a larger scale, Weather Underground (www.wunderground.com) monitors more than 10,000 U.S. and 3,000 worldwide personal weather stations plus thousands of weather webcams to get an extremely accurate picture of weather conditions everywhere.

In addition to dozens of websites that provide location and destination maps, travel websites such as Tripcheck.com and Traffic411.com show you the road conditions along your route. Many people use these every day to plan their commute based on information displayed on a live-action map of their route and traffic cameras that show what's actually happening on the road at that moment. These sites can also prove valuable if you need to plan an evacuation route, showing road conditions, traffic hazards, and recommended routes. Satellite photography such as Google Earth is good for accurate maps, but most of them are not up to date and may not show you current conditions.

GPS systems are very inexpensive these days. Many people buy services through their cell phone providers that display a map and also talk to you about where and when to turn and how far to go, similar to having someone with a map sitting in your passenger seat. GPS systems are great for regular commuting, long-distance road trips, and finding your way through an unfamiliar city. Unfortunately, they are not only unreliable in a disaster situation, they can actually endanger your life.

A GPS system tells you where you are and how to get wherever you're going. But if road conditions have changed, the GPS won't tell you. For example, if you're driving on a highway, the GPS may tell you to take a certain exit. What the GPS doesn't know is that an hour before, the exit was changed to an entrance because the traffic load from an evacuation route demanded added capacity. There've also been cases where federal disaster workers were relying on the GPS and almost drove off the edge of a ramp that no longer had a bridge deck. Because you tend to believe whatever the talking box tells you, you're lulled into a false sense of security when you follow the directions.

Cell phones are so common that most people don't think of them as "new" technology, but they can be used in new and interesting ways that make them more useful in a crisis. Apart from standard text messaging and person-to-person, walkie-talkie style communications, there are mass-communication systems that can reach hundreds of thousands of people simultaneously. One of the first applications of this is the development of an early alert system for school administrators to identify when a crisis is imminent. Some schools have similar systems, where all students on campus register their cell phones and the school can then send alerts specifically to the students.

Other uses of technology include:

- Voice Over Internet Protocol (VOIP) for Internet-based phone or communications where regular phone lines are down or unavailable.

- A manually powered emergency mobile phone charger. Cranking the handle for a few minutes gives you enough juice in the cell phone to make a short call.

- "Shake" flashlights that use LEDs to provide a high-intensity, long-lasting light.

◆ Kid finders, remote locator devices that help you find a lost child or other lost member of your tribe.

◆ Satellite personal tracking devices that transmit your GPS coordinates every few minutes to assist rescuers to locate you.

Social Networking and Mass Communications

One of the latest things in social networking is micro-blogging. Micro-blogging is similar to blogging, except that you can only send brief text updates (typically 140 characters or less) and micromedia in the form of small photos and audio clips. Micro-blogs combine the immediacy of instant messaging with the public availability of information on a blog. The content tends to be very topical.

Twitter is currently the best-known micro-blog, but there are many others, including micro-blogs off of better-known social networking sites such as LinkedIn, MySpace, and Facebook. The Red Cross and other organizations are using micro-blogs as a way to push data out to people about disaster conditions and so on. It's a no-cost method: anyone who logs in can look.

Now that you've learned about the basics of needs-based planning, you can build a disaster plan for you and your family. Specific information on how to create a disaster plan for you and your family appears in Appendix C.

The Least You Need to Know

◆ Most people think (incorrectly) of good disaster planning as being for specific types of disasters.

◆ Needs-based planning is the preventative medicine of disaster. You can't prevent the disaster, but you can prevent it from having an effect.

◆ It's okay to withhold information during a crisis, but don't lie.

◆ Never let your business procedures be dependent on a single person or system.

◆ Technology can be helpful, but don't rely on it unthinkingly. In some cases, low-tech solutions may be safer and more reliable.

3

Packing a Personal Go-Pak

In This Chapter

- ◆ What goes into a Go-Pak
- ◆ Packing safety and sustenance supplies
- ◆ Keeping in contact
- ◆ What kinds of documentation and financial information to take
- ◆ Special considerations when packing for children

The Boy Scout motto, "Be prepared," is never truer than when it comes to disasters. Years of experience with survivors who were prepared—and those who were not—shows that the most important thing anyone can do to be prepared is to have a Go-Pak.

In this chapter, I describe a basic disaster Go-Pak that you can assemble yourself, and compare this homemade kit to several commercially available kits. Finally, I describe a project you can do with your children at home or in school to help them be prepared with a Go-Pak of their own.

What Is a Go-Pak?

The basic disaster Go-Pak must contain items designed to assist you in staying safe, keeping you sustained until other assistance can arrive, keeping you in contact with family and friends, and ensuring that your important documents are available when you need them. Unlike other chapters in this book, this chapter is designed as a packing list with an overview of why each component is so important.

The materials in your Go-Pak should be functional, durable, and as lightweight as practical. You don't need to go to a sporting goods store and buy an expensive new version of everything in the following lists. You probably have most of the items already, and most of the rest are available at the drug store or a general retail outlet.

You can build your own Go-Pak or buy one ready-made. Building your own Go-Pak is as easy as buying a backpack and filling it. Clothing is best packed using Space Bags or a similar packing and storage bag that allows you to suck or squeeze the air out and keep the clothing clean and dry. Similarly, you should keep liquid toiletries in zip closure bags. (It's not a bad idea to use a second bag as insurance that nothing leaks from a spill in the first.) Finally, all documents should be in a third waterproof bag.

Commercial Go-Paks have the advantage of being preassembled and thus much easier to prepare. These kits contain largely the same items as a do-it-yourself Go-Pak. The main difference is that as you assemble your own Go-Pak, you learn what's in it. A commercial kit contains generic toiletries and all the safety and survival items possible. You must add in your own clothes, but a good commercial kit comes with bags to ensure the safety of your clothing and documents.

You need to pack several categories of items in your Go-Pak: safety supplies, sustenance supplies, communication supplies, and information and important papers.

The following lists show the supplies you need for one person. If you're preparing Go-Paks for several people, consider having a separate Go-Pak for each person.

Safety Supplies

The first category of supplies you need to pack in your Go-Pak are safety supplies. Safety supplies are the things you need to walk out of a location and survive for the next day or two. They're literally "the clothes on your back" plus some basic survival equipment.

Personal Clothing

The personal clothing you pack should be durable and functional. You don't need to spend a lot of money on clothes in your Go-Pak; in fact, you'll probably feel better if your survival clothes aren't particularly fancy. You don't want to be wearing clothes that you worry about snagging or getting dirty if you have to move fast or work hard.

In general, clothing made of natural fibers is better than clothing made of synthetic fibers. It's more durable, easier to repair if it rips, and more comfortable in extreme conditions. Natural fibers are a little easier to wash in field conditions, which can be good if you're stuck in a shelter or on the road for an extended period. Here is a list of necessary personal clothing:

- ❑ Hat (to prevent sunburn and provide shade, and to keep the rain out of your face)
- ❑ Closed-toe shoes (to protect your toes)
- ❑ Jacket with hood
- ❑ Poncho (or rain coat and pants)
- ❑ Long trousers, one pair (to protect your legs)
- ❑ Shorts, one pair (cooler for hot weather)
- ❑ Long-sleeved shirt, one (to protect arms and prevent sunburn)
- ❑ Short-sleeved shirts, three (cooler for hot weather)
- ❑ Underwear, three or more
- ❑ Socks, three pairs or more
- ❑ Large bandanna, one or more (to prevent sunburn)

Miscellaneous Items

The miscellaneous items are the minimum basic tools you should have on hand for survival purposes:

- ❑ Lock and key (to keep the Go-Pak locked when you're not using it)
- ❑ Personal first-aid kit
- ❑ Extra sets of car and house keys
- ❑ Knife, lockable
- ❑ Multi-use tool (such as a Leatherman or Gerber)
- ❑ Work gloves (to protect your hands during clean-up or heavy work)
- ❑ Face mask or dust mask (to protect your lungs during clean-up or heavy work)
- ❑ Flashlight and batteries
- ❑ Whistle (to signal and call for help)
- ❑ Compass

Sustenance Supplies

With the basic survival supplies taken care of, you need to look at packing sustenance supplies. Sustenance supplies are the necessities for staying alive without access to any outside services or facilities, the personal items that make life more pleasant (such as soap), a supply of prescriptions, and basic medicine cabinet supplies (such as aspirin and antacids). They'll make living away from home much more endurable and nicer for everyone around you, too.

Personal Gear

The personal gear in this category is mostly toiletries. You could live without all these, but you're likely to feel a lot better with brushed teeth and clean, brushed hair:

- ❏ Waterproof watch, preferably with an alarm
- ❏ Sunglasses
- ❏ Extra set of prescription glasses
- ❏ Personal hygiene items
- ❏ Bar soap
- ❏ Shampoo
- ❏ Razor blades
- ❏ Shaving cream
- ❏ Deodorant, unscented
- ❏ Toothbrush, toothpaste, dental floss
- ❏ Comb, brush, unbreakable mirror
- ❏ Toilet paper
- ❏ Lip balm
- ❏ Hand lotion, unscented
- ❏ Sunscreen, unscented
- ❏ Insect repellent—pump type, unscented
- ❏ Flip-flops or some other shower shoe
- ❏ Biodegradable laundry detergent
- ❏ Anti-fungal foot powder and moleskin (for blisters)

Life Preservers

Always use unscented toiletries to protect others at the shelter from allergic reactions and asthma attacks. In addition, shelters are usually very close quarters. Not dousing yourself with scents is the considerate thing to do.

It's not essential, but you may want to avoid pressurized cans of toiletries. They're bulky and they can be a fire hazard if near an open flame. Use things such as stick or roll-on deodorant and tube or pump shaving cream.

Medications

Medications include a two-week supply of prescriptions that you or anyone in your family needs, as well as general OTC medications that make life a little easier:

- ❏ Prescriptions (a two-week supply at least)
- ❏ Aspirin
- ❏ Antacids (Maalox/Mylanta/TUMS)
- ❏ Antihistamines (Benadryl or Zyrtec)
- ❏ Tylenol
- ❏ Decongestants
- ❏ Imodium (anti-diarrheal medication)

Miscellaneous Items

In addition to toiletries and medications, you'll also want to pack a number of additional supplies for staying away from home and normal services for an extended period:

- ❏ Towels (large and small)
- ❏ Washcloth
- ❏ Books, reading material, crosswords, Sudoku
- ❏ Inflatable pillow
- ❏ Handi-wipes, unscented
- ❏ Laundry bag and clothespins
- ❏ Safety pins and sewing kit
- ❏ Waterproof matches and fire starter material (no flammable fluids!)
- ❏ Electrical tape
- ❏ Parachute rope, 20 feet
- ❏ Extra shoelaces

❏ Water-purification tablets

❏ About 72 hours' worth of high-nutrition snacks and pocket snacks (such as trail mix, gum, dried fruit, and hard candy)

❏ Notebook, pens, pencils

❏ Cup (pocket size, collapsible)

Communication Supplies

You may want to pack a number of communication supplies. At a minimum, you need the following:

❏ Small portable radio with headphones

❏ Roll of quarters for a payphone

❏ Cell phone charger (12-volt car and 110-volt wall outlet)

You may also want to get walkie-talkies, which are available at many retail outlets. (Some cell phones offer walkie-talkie style operation, but they may depend on the cell system being in operation.)

Prep Facts

Ham radio operators are licensed to use fairly powerful—and very compact—handheld units that can be used as walkie-talkies from person to person or can be used to contact other people. Getting a ham radio license is fairly easy, although it does require some study. For more information about ham radio and how to get a license, look at the American Radio Relay League (ARRL) website at www.arrl.org.

Information and Important Papers

The materials in the preceding three categories are all aimed at keeping you alive and in some measure of comfort while you're living away from home. This category, information and important papers, is designed to help you maintain and recover your normal life as quickly and smoothly as possible.

Carry the following documents with you or in your Go-Pak at all times when you've evacuated and are on the road or in a shelter. (The papers you need to carry with you are relatively compact and don't take a lot of space in the Go-Pak.)

❏ Driver's license

❏ Insurance policy documents

❏ Medical insurance cards and information

❏ Vehicle and real estate titles

❏ Birth certificates, Social Security cards, and passports

❏ Credit cards

Keep photocopies of all these documents in a safe deposit box as well. Although these may not have the same force as originals, they will provide a good record that will help you track back to the duplicates. Moreover, safe-deposit boxes tend to be pretty secure, regardless of the type of disaster. Carry the safe deposit keys with you.

If you have stocks or bonds, you can keep the originals in the safe-deposit box but carry photocopies of them with you. In a pinch, it is possible to request a replacement stock certificate for a small fee, allowing you to sell the stock even without the certificate in your physical possession.

You should also have copies of several sets of records that may be difficult to replace. (Keep copies of all these records in the same safe-deposit box as your ID and other information.)

❏ School records for everyone in the family who's in school. (In a major disaster, it's possible that the school and records will be destroyed.)

❏ Medical and immunization records for everyone in the family. It's likely that copies of these will eventually be available, but if someone needs emergency medical care while you're evacuated, having medical records handy may save someone's life. Medical records are especially important if you need to prove that you have to carry syringes, equipment, or some types of medication.

❏ Mortgage, boat, auto, and personal loan information, including ID number, monthly payment amount, company contact information, and address. (If the mortgage or loan is in one person's name, make sure you get something set up with the lender company that allows another authorized person to talk about the account even if the primary person's not there.)

❏ Credit card numbers, including exact name on the card, expiration date, authorization code, company contact numbers, and monthly due date. You can put a password on your accounts to make it harder for someone to use your card. Do that, but don't write the access passwords down anywhere.

❏ Bank and brokerage account numbers and the branch names and contact numbers.

❏ Marriage certificates and custody documents. If someone in your family needs emergency medical care or there's a question about reclaiming your children at a shelter, these documents can smooth the way by proving that you're authorized to do these things.

In addition, everyone in your family or evacuation group should get a copy of emergency contact phone numbers for your friends and family members, as well as policy numbers, agent names, and the 24-hour emergency contact numbers for medical, home, car, boat, flood, and life insurance. Everyone in the group should also have a copy of the emergency plan.

Having all your documents with you in one place is also good prevention for someone stealing them from your home while you're out and then stealing your identity using them. When you evacuate, keep copies of all these documents hidden in your house in a waterproof container. If you lose the documents with you, you may be able to recover the set from your house even if the bank branch with your safe-deposit box is closed for a while.

Life Preservers

You may want to carry copies of your documents and key information on a flash drive. Having financial, medical, and school information in an electronic form as well as printed can make it easier to copy or transmit. You can also e-mail electronic files to yourself so they're waiting for you whenever you're able to log on from any location.

Other Financial Concerns

When you're evacuating, you should have cash, your checkbook, and credit cards with you. It's a good idea to have perhaps as much as a few hundred dollars with you, including small bills and change.

Depending on how bad the disaster is, don't expect to be able to do much with a checkbook. Also don't expect to write checks drawn on local banks within the disaster area.

Additional Considerations for Children

In addition to the materials described in the preceding lists, Go-Paks for children should also include photos of family members with the names written on the back, and the names and contact information for at least four family members and friends, which they should memorize. In addition, children should have a list of the local information sources, such as TV and radio stations, weather services, local access cable, local government emergency operation center (EOC) numbers, and local print media.

The Least You Need to Know

- Every family member, friend, employee, and client should own a Go-Pak.

- The Go-Pak is designed to provide 72 hours of self sufficiency.

- Disaster Go-Paks make the perfect holiday gift: the gift of preparedness.

- Children as young as five can be taught how to build and maintain their own Go-Paks.

- Your Go-Pak goes wherever you do.

Chapter 4

Special Needs

In This Chapter

- ◆ Getting information about shelters that accommodate special needs
- ◆ How children deal with stress
- ◆ Issues that seniors may face
- ◆ Sheltering with service animals and pets
- ◆ Dealing with the physically and mentally disabled
- ◆ Shelter for the homeless

Disaster planning, rescue, and recovery have come from almost nothing to being pretty sophisticated in the last 50 years. But the one group that has consistently been hard to save is people with special needs and disabilities, including children, seniors, physically or mentally challenged people, and the homeless. If you don't need to evacuate, your life will probably be relatively unaffected, but if you have to take shelter elsewhere, things can become far more difficult. Although people with special needs by definition have additional problems or issues, you can do a great many things to accommodate them.

In this chapter, I tell you how to get information about shelters that accommodate special needs and how to deal with the problems faced by children, seniors, the disabled, and the homeless. Pets and service animals present many of the same difficulties and challenges as people with special needs, so there's a section on what to do to make sure they weather the disaster safely, too.

Getting Information

The normal routes for getting information about shelters are not nearly as effective when you're dealing with special needs requirements. Public warning systems don't always mention that there are shelters at locations. It's always a good idea to know where your local shelters are in any case, but if you or a family member will have special requirements, it's imperative that you know which shelters are likely to be most accommodating.

News

You may know where the shelters are in your area, but what if you're out of your area on business or vacation somewhere? If you don't know the location of local shelters equipped for special needs, your best bet for information is TV news. When there's an evacuation, news programs announce locations of shelters and tend to announce which shelters have additional facilities or options.

Public Warning Systems

Public warning systems are set up for people who can see and hear. But what do you do if you're blind? There are emergency weather radio units that wake up when there's an emergency broadcast. (Actually, many people living in tornado zones have units similar to this in their house.) When the units wake up, there's an audible alarm followed by announcements.

Sirens and alarms work great if you're blind, but not if you're deaf. However, if you are deaf and you live alone, you can get a *siren dog*. There are also emergency weather stations that flash a light and then display text on a screen or TTY unit. Most TV-based emergency broadcasting systems use both scrolling text and a voiceover.

def•i•ni•tion

A **siren dog** is a service dog trained to respond to sirens, smoke alarms, and alarm bells by crawling up on their owner, licking the owner's face, and tugging at clothes and hands.

Kids Can Be Fragile

During an evacuation, adults are going to be stressed. They can be tense, snappy, depressed, or a hundred other things, but adults have the life tools, experience, and vocabulary to talk about what they're feeling and why. It may not be possible to resolve how they're feeling, but their feelings can be expressed and discussed.

Kids are just as likely to be impacted by the stress of evacuation as an adult, but they don't have the same ways to deal with how they're feeling. Depending on how old they are, children may regress emotionally or they may even get older.

Watch for Behaviors

When placed under significant pressures, children who are younger than seven or eight become more infantile. You're likely to see bedwetting and separation anxiety. They may even have a complete alteration of their personality: the kid who's very adventurous may be the kid who never leaves Mommy's leg, or vice versa. They're trying to address the stress and their feelings.

In contrast, grade school and middle school kids will "grow" and assume responsibilities. If you have a precocious third-grader who helps deliver food, that's normal. However, a typical middle child who's now suddenly acting paternal and even disciplining their older sibs is a warning sign.

What's happening is that the child is mimicking the people they normally see. They may also mimic other adult behaviors, such as not asking for help or mimicking the behavior that they perceive: "Mommy always comforts me, but I don't comfort Mommy." They can shut off their own abilities to process.

Adolescents and young adults under stress will go in one of two directions. The first possibility is that they'll go into brooding teenager mode, even if they were very outgoing to begin with. The other possibility is that they'll go into "very adult" mode. Teenagers tend to explore and take risks, and this manifests as perceived adult risk-taking behaviors. This is not just that they're the shelter volunteer or helping out in other ways. It'll be a marked behavior change. For example, a 13-year-old might become precociously sexual, coming on to adult males or females. They may become the teen who gets into drugs or alcohol to deal with the stress. Teens are notoriously bad at delayed gratification, which can easily result in them making poor decisions.

Storm Warnings

Seeing significant behavior or personality changes in a child is a major warning sign. It means that the stress they're experiencing has completely overwhelmed their ability to compensate internally. It's important for parents to realize that all this acting out is not a sign of weakness in the child, nor of poor parenting. It's just the way kids deal with huge amounts of stress.

How long behavior and personality changes take to manifest depends on the child and how severe the impact is, based on the proximity, intensity, and duration of the stress. (The bigger the exposure, the bigger the shock.) But it's important to note that this could happen very early. It all depends.

Help Them Process What's Happening

You can do many things to help children deal with the stresses of the situation. First, maintain a level of normalcy. Keep a set of normal processes that the children are used to, such as brushing teeth before bed and making their beds in the morning. If regular school facilities aren't available, consider establishing a home school and having classes even before the storm begins. All these things establish the idea that, although you're in a shelter and things are definitely not normal, you are in control of the structure of your life.

When you're packing Go-Paks or evacuating, make a point of bringing games or stories for the kids. Keeping the kids going with crayons, puzzles, cards, or Monopoly may be one of the best ways to help them cope with a difficult situation. It will also give you something to do to while away the time: shelter life can be dull.

Fantasize the problem with the child. You can say "I'm excited" or "I'm scared" and let the child take it from there and see what they have to say. But you give them the fantasy power to deal with it. Kids do best when they're allowed to express the problem in their terms. Many parents are afraid to say, "That was very scary about the quake; you must be scared to go home." Go ahead and say it. The kids already know it's scary; after all, they're at the shelter.

> **Prep Facts**
>
> Shelters frequently have counselors to talk to kids about how they feel. Check with the shelter manager or medical team to find out what's available.

You don't need to explore every single feeling, but you should provide open-ended questions, such as, "We're going to go home soon. How do you feel about that?" rather than "Are you scared?" which equates to "You should be scared." For example, suppose there's a crack in the plaster after a quake. Ask the child how they want to deal with this. The child might say "I've always wanted a Care Bear poster." Go out and get a poster. The big thing is to give the child the power to deal with this.

The best thing you can do for children after a disaster is to re-establish a sense of normalcy and routine. Adults feel this way, too: it's very common after a disaster for adults to want to dig in and fix things. But children just want a routine, because it means that things are predictable. Getting the schools open and getting the kids back into their routines makes them feel safe, which makes their parents feel better, too.

Seniors Can Be Fragile, Too

Even though seniors don't require the majority of medical attention in disasters, they do have some special needs that need to be taken into account.

There are three classes of seniors:

◆ Seniors who are just older. The vast majority of seniors aren't broken, they're just older. They move a little slower and may be a little more fragile, but that's it.

◆ Seniors who are mentally intact but are facing medical challenges. This group requires things such as oxygen, nebulizers, walkers or wheelchairs, or some other assistive device. They're medically needy.

◆ Seniors who may or may not be physically disabled, but they're not mentally there for whatever reason, such as Alzheimer's patients or stroke victims.

Planning's Even More Important!

Disaster planning is extremely important for seniors, particularly those with physical disabilities. They have to plan for a caregiver as well as whatever medical equipment they may require. For example, if you need to use an oxygen concentrator, you may need a plug in the floor at the shelter.

If you're using a home health agency, chances are good that they'll have a disaster plan. Check to find out what services the agency may offer in case of emergency.

Packing Two Weeks of Meds

There's an old saying that you can tell how old someone is by the size of their medicine cabinet. Most seniors need a regular supply of prescription meds. As part of your disaster planning, you should have a two-week supply of meds in your Go-Pak. (If your meds are particularly essential, such as heart medicine to prevent fibrillation or anti-rejection meds, consider having a four-week supply in your Go-Pak.)

You can build up a supply for your Go-Pak fairly easily. For example, if you're getting your prescriptions filled by mail for 90 days at a time, also have a 30-day prescription filled locally as soon as you get the notice that the 90-day prescription's in the mail. When the 90-day supply arrives, fill up the 30-day bottles and put them in the Go-Pak. Be sure to swap out the meds in the Go-Pak every 6 to 12 months to keep them fresh.

Caretaking Requirements

All shelters require that you have a caretaker if you have medical requirements. If you haven't got one, you aren't going to be turned away, but you'll have to share the caretakers on staff with whoever else needs a caretaker.

Storm Warnings

Current statistics show that half of all adult Americans can't read a prescription drug label or find a specific piece of information in a short publication. As a result, directions, warnings, or instructions for disaster alerts and sheltering won't reach their intended audience. (This also applies to people who don't read English.) Many communities have been expanding their disaster preparedness materials to include more graphics and visually based materials, but this is a slow process. The illiterate are likely to feel increased stress because their normal support systems aren't available. If you are dealing with someone who you think may be having trouble with reading instructions, take a moment to go over directions orally with them and make sure they understand.

Service Animals and Pets

Service animals include seeing eye dogs, hearing ear dogs, seizure dogs, siren dogs, and psychiatric service dogs. (There are other types of service animals, such as assistance monkeys.) There are also search-and-rescue dogs and cadaver dogs, both of which are used primarily by police and rescue organizations.

Prep Facts

Service animals are for people with some physical or mental condition that prevents them from functioning effectively. The service animal has special training that lets them mitigate their owner's disability. Owners have a right to service animals in most public places and shelters because the dogs allow them to function and cope with life in an unfamiliar location. In contrast, emotional support animals are important, but they're not truly service animals. They don't have to be trained (beyond basic animal training). All shelters will accept service animals, but not all will accept other animals.

Pets—including emotional support animals—can include dogs, cats, birds, snakes, horses, all the way up to lions, tigers, and bears.

After Katrina, a number of locations have municipal laws that part of disaster preparedness will include pet-friendly shelters. This doesn't mean there'll be enough slots, though, nor does it necessarily mean that a pet-friendly shelter will be able to take your pet. And no matter what, it sure won't be similar to living at home.

Food

Although the shelter may be pet-friendly and willing to allow your animal in, feeding your animal is not part of the deal. Some shelters will provide food, thanks to a local donor or a pet food manufacturer, but the shelter isn't likely to have anything stranger than dog and cat food. In addition, you'll probably be required to do the feeding. If you have a pet that requires any special food, you should bring it with you.

Shelter

A dog or cat in a carrier is okay. A bird in a cage is okay. A lion on a leash is not. A snake in an appropriate snake container is okay, but a boa constrictor around your shoulders isn't. And just because you love your pet, don't expect everyone to feel the same way. A chopper pilot in an extreme rescue situation may not let you on with a loose snake. But no rescuer would deny you access if you've put your pet in an appropriate cage.

Sheltering your animal protects it from serious injury or death and may also save the life of people in the area. During Hurricane Ike, there were a couple of tigers running loose on the island for up to a week. No provisions had been made for feeding the tigers; they had just been released from their cages by their owner. (The emergency teams were not initially aware of this and they were not pleased when they found out about it.) But even abandoning a dog or a cat is grossly irresponsible: after every evacuation, many animals have to be shot because they were seriously injured when they were found.

As you're making plans, it's important for you to remember that the shelter is not a kennel, it's simply a location. It's not going to be a lot of fun. You're going to be expected to take care of cleaning, feeding, walking, and entertaining. If the wind's howling, though, you're not going to be able to do anything else anyway except clean the crate your pet's in.

Toys and Supplies

As with children, it's a good idea to bring toys and other supplies for your pets. Bring a favorite chew toy or rawhide bones for dogs, stuffed mice or cat-fishing poles for cats. For rabbits, you need gnawing bones, hamsters should have wheels, and so on. This will help keep them occupied and restore a little normalcy to their lives, too.

Prep Facts
If you consider your animal a member of the family, it should have its own Go-Pak. Animals should also have their own crate or container, which should be part of their Go-Paks.

Disabilities Can Affect Your Planning

Planning for people with physical or mental disabilities has a lot of overlap with planning for seniors. Because of the number of extra problems or conditions you have to cope with, you should make a strong effort to plan and prepare if you or a family member is disabled.

Caretakers

If your health is compromised to the point that you require a caretaker, you must include the caretaker in your evacuation plans. Public caretaking at shelters is going to be spotty at best, with as many as several hundred people for each nurse or practitioner. Moreover, the medical staff at the shelter won't be immediately familiar with your medical needs and, in a real crunch, may not be adept at the procedures you need to survive.

People with mental disabilities, including retardation, dissociative disorders, or behavioral/cognitive dysfunction problems, may not have caretakers. Under the stress of being in the shelter, their responses are generally going to be an exacerbation of their problems. For example, if it's someone with a cleanliness fetish, you may end up with the cleanest shelter around. But there is also a chance that the person may have a paranoid break and become a danger to themselves and others.

Shelter counselors are trained to help individuals who are in the early stages of decompensation and can frequently talk them down and help them re-integrate. If that doesn't work, it may be possible to evacuate them to a safer place where they can get medication and further treatment. However, the ultimate responsibility is the overall safety of the shelter. If someone is becoming hysterical because they're feeling trapped in the shelter, and the staff can't talk them down, it may be necessary to isolate the person or strap them down so they can't hurt themselves or panic the rest of the shelter. The options may be limited but it's the best that can be done. (It's a lot like what happens to unruly passengers in airplanes who refuse to calm down: if necessary, they're subdued until the plane lands.)

Wheelchairs and Other Adaptive Equipment

You can count on shelters being wheelchair-accessible. However, there may not be access to electricity. If the shelter is running on emergency power, charging battery packs on powered wheelchairs is way down the list of allowed uses. Most assistive devices have a manual mode, but you won't be getting around as quickly or as easily. Do take your devices with you if you need them.

Label everything! Walkers, canes, chairs, and even leg braces look like everyone else's. Write your name with a permanent marker on walkers and other equipment and then cover that with clear tape so it doesn't rub off.

Glasses are famous for looking alike. A colored piece of tape on the earpiece is probably adequate.

Transportation Needs

If you have special transportation needs in your day-to-day life, you'll need to make plans for how you're going to evacuate. Many medical transportation firms have contingency plans for evacuating special needs clients, but it's also very possible that they won't be able to. (For example, a quake or flash flood could have closed the roads.) If possible, have a back-up plan in place, such as using your own van driven by a friend or making transport arrangements with a friend.

Planning for Service Animals

Service animals, by law, are allowed to be by the bedside of the person, but you have to go to a pet-friendly shelter. Your service animal must also have a service animal vest on. This is also a symbol to kids not to touch or play.

Dealing with the Homeless

Most shelters open to adults are also open to the homeless. They may even provide special needs assistance to the homeless. For example, one of the issues is that, if you're homeless, you can't get sent back to a home on streets that don't exist anymore. As a result, many shelters have social services for people who are displaced by the disaster.

If you're helping homeless people take shelter, you'll need to locate shelters that may offer these services. You should get a guide from the local emergency management office, the Red Cross, or maybe even the news. But do this as early in the planning process as you can—as soon as you know you need these services, not when the storm's looming on the horizon.

If a disaster is forecastable (such as a hurricane), the news will also talk about at least some of the services available at shelters. There will also be information about shelters specifically designed for homeless people. In addition, hospitals and schools provide information on special needs services.

Homeless individuals tend to be familiar with shelters when there's a reason to be there, so they'll tend to congregate there in a disaster, particularly if the shelter they normally go to is also a disaster shelter. However, most homeless shelters aren't equipped to be disaster shelters, because they lack back-up power, services, and a safe locale.

Some shelters are set up with absolutely everything: medical volunteers, social workers, and a way to deal with the homeless afterward. Shelters are the great equalizers: you can have people from million-dollar homes sleeping next to people who clean million-dollar homes as well as people who live under a bridge.

Most of the problems in dealing with the homeless in shelters are the attitudes of others. There aren't a lot of communicable diseases. There isn't going to be a lot of panhandling or mooching because that's taboo in shelters. Moreover, there are strong territorial rules among most homeless people: the kindergarten rule of "If it ain't yours, don't touch it!" applies here. The homeless will look out for each other's possessions, and your things are usually at less risk from the homeless than they are from other people. The biggest problem with homeless people in shelters is getting them out at the end of the disaster, simply because there isn't a place for them to go. There may be more prodding needed.

The Least You Need to Know

- ◆ "Special needs" includes people younger than 18 or older than 55, the physically or mentally challenged, people with chronic illnesses, the illiterate, the homeless, and service animals and pets.

- ◆ Evacuating is stressful for everybody, but special needs populations have fewer ways to address the stress and to cope.

- ◆ People with special needs take longer to rescue, longer to evacuate, and longer to shelter.

- ◆ People with special needs suffer more stress in a disaster specifically because of their requirements, resulting in more pronounced stress behaviors.

- ◆ How you react to stress is personal. What's shocking for you may be routine for others.

Chapter 5

Safety Isn't Just Good Preparation

In This Chapter

- ♦ Determining when sheltering in place is appropriate or necessary
- ♦ Planning for an evacuation
- ♦ Dealing with life in a shelter
- ♦ Returning home and getting back to normal

You can have all the supplies you can carry, but it won't make a lot of difference if you don't have a safe place to take shelter. There are lots of options for taking cover, some better than others. Depending on where you are when disaster strikes, you may have to improvise, but knowing what to do may help you make better choices.

In this chapter, I provide you with the tools to prepare yourself whether you will shelter in place or evacuate. Then I help you gather the resources you need to shelter in place as well as the ability to tell when it is appropriate to shelter in place and when it is not. For those times when evacuation is a must, I show you how to plan your egress and become an integral part of shelter life.

Sheltering in Place

Sheltering in place means just that. You're taking shelter from the disaster wherever you are: in your car, your home, your business, or another impromptu location such as a shopping mall.

There are several reasons to shelter in place. A disaster can occur so quickly that there's no advance warning and you do the best you can. Earthquakes and toxic spills strike without warning, of course, but other disasters such as tornadoes or flash floods can happen so suddenly that there's no chance to prepare.

The other reason to shelter in place is that there's no safe place to go, even when you've had plenty of warning. For example, you may know that a hurricane is coming for several days, but the area affected by the hurricane may be so large that it's impractical to try to escape. Similarly, floods or wildfires may have cut off evacuation routes or the amount of traffic on the roads makes it unlikely that you'd be able to reach shelter. You may also be unable to evacuate for other reasons, such as disability or inadequate transportation. Even when evacuation to a safer place is possible, people frequently choose to shelter in place because of emotional attachments.

What Shelter Is Right?

The situation will determine the kind of sheltering in place to take. If a train derailed and released a tank car of chlorine gas, there may not be time to get out of the area, particularly if you don't know which direction is away from the spill. In that case, the safest place to be is behind several doors with the air conditioning turned off. If you're in danger from a tornado, your safest place is in a sub-basement. On the other hand, if you're driving and you see a tornado, you're safer getting off the road and into a ditch than staying in the car.

Life Preservers

Your best chance of survival is to go to the closest place that provides the maximum amount of shelter.

When there's a predictable hazard with a sufficient lead time, head for a shelter that provides you with the maximum amount of protection. For a Category 1 hurricane, if your home is designed to take 125 m.p.h. winds and you have back-up power in the form of a generator, you might as well stay. On the other hand, if you're in the path of a Category 5 hurricane, it's time to get out.

Food, Water, and Air

You need 72 hours of food and water for each person taking shelter. You need a bare minimum of a quart of water per person per day, but that's just for drinking and nothing else: no washing, no teeth brushing, nothing else. For comfort and safety, you should store at least half a gallon per person per day, and even more if you have the ability.

Bottled water is available in 2.5 gallon plastic containers. These are easy to store and use. Check the shelf life for the water and rotate the water supplies so the water isn't outdated. (The water may not go bad in terms of infection or mold, but it may absorb a plastic taste from the jugs and taste bad.) Keep water jugs stored in a dark place and away from chemicals and things with strong scents such as laundry detergents.

One of the reasons that stored water tastes flat is that there's no air dissolved in it. If the taste of stored water bothers you, fill a jug about halfway, then cover it tightly and shake it for a minute to dissolve some of the air. It should taste a little better afterward.

If you're not sure about the quality of the water you've stored, you should purify it. The best way is to boil it for a minute or two (three to five minutes if you're above 5,000 feet), but this may not be practical. You should store water-purification chemicals as part of your standard emergency supplies to disinfect water that may be impure. Most water-purification tablets available today use chlorine or iodine to kill the bugs that might be there.

If you didn't stock water-purification tablets, you can use 8 drops of unscented liquid chlorine bleach per gallon (or 16 if the water is cloudy or murky). Add the bleach, slosh a little of the water on the container's neck and cap to clean them off, let the water stand for 10 minutes, and then pour it into a sanitized container. You can sanitize water containers with more bleach: use a sanitizing solution of a tablespoon of bleach per gallon of water. If the water has particulates or debris, filter it through a couple of coffee filters or napkins.

In a real pinch, you can sterilize water with standard tincture of iodine from your first aid kit. Use 5 drops of tincture of iodine per quart of clear water (10 if it's murky) and let it stand for at least 30 minutes. The water will taste a little funny, but it'll be okay. *Don't* use mercurochrome—it's poisonous.

Prep Facts

In an emergency, you can use beer as a substitute for bottled water. Beer also has the advantages of being pasteurized, handily packaged, and readily available.

Although it's true that alcohol dehydrates you, beers with 3.2 percent alcohol or less (most name-brand American beers and virtually all "lite" beers) provide just enough water above the dehydration effect of the alcohol that you'll stay hydrated (and delightfully buzzed). Imports, higher alcohol beers, malt liquors, and heavier beers have too much alcohol and malt and will dry you out. The average adult needs 6 to 10 (12-oz.) beers a day to stay hydrated.

Remember that you're going to be a bit drunk, so don't drive and be careful. On the other hand, there are worse ways to spend a disaster than by staying inactive and drinking a lot of beer.

In addition to water, you need food. You don't need to buy expensive, freeze-dried, foil-packed food, nor do you need to have military-style MRE (meals ready-to-eat) packages, but you do need some kind of food that stores easily and doesn't necessarily require a lot of cooking.

Canned goods are an excellent way to get a variety of food. Canned food stores well, has a long shelf life, and is easy to access. Furthermore, although most canned food tastes better when heated, it doesn't actually require cooking if you're in an emergency situation. Plan on storing a selection of canned vegetables, fruits, and prepared foods such as chili, stew, or soups. Dried nuts, beans, and pasta may be good, too, but remember that they'll require you to boil adequate quantities of water, which may not be possible depending on the nature of the disaster and your shelter. Don't forget to keep a few can openers with your supplies, too: it'd be silly to have a week of food in front of you and no way to open the cans.

You also need to be able to control the ventilation where you're sheltering. There should be both active and passive ventilation so you don't suffocate. For most hazards, air quality isn't a problem, but for things such as wildfires or toxic spills, you need to be able to shut off the outside air. For the most part, you can do this by shutting off your air conditioning or heating so outside air isn't getting pumped into the house and blocking the vents. There's enough air in a house or apartment to keep you safe for quite a while; even a single room will have enough air for a couple hours. For longer-term shelter, you may need a filtering system to remove smoke or other gasses from the outside air.

Building a Safe Room

You may want to consider building a safe room for general disaster shelter. Safe rooms can be anything from a simple storm cellar or underground shelter with a door that you can bar from the inside, to concrete bunkers reminiscent of home civil-defense shelters with self-contained power, supplies, and metal doors. You can also buy pre-fabricated safe rooms that are basically large steel boxes.

> **Storm Warnings**
>
> You won't have a lot of room in a safe room, but they're all about safety, not comfort.

Before taking shelter in a location, you need to be sure that it's going to provide protection from the disaster you're facing. A storm cellar may be protection from a tornado, but may not be any good if you get flooded. An upstairs room that provides shelter from flooding could trap you if you're sheltering from a wildfire.

The Evacuation

Evacuation deals with leaving your home or your current location. It costs you emotional connections. Your home is your normal point of safety and security where your possessions are stored. When you evacuate, you have to acknowledge that your current situation isn't nearly as safe as the place you're going to.

Gassed Up and Ready to Go

Regardless of the type of disaster, you need to be ready to leave at a moment's notice. Because disasters can strike at any time, you need to be ready all the time. This doesn't mean that you need to sit by the door all day, quivering in anticipation with a news channel on for announcements, but you should have your Go-Paks by the door or with you in the car and your gas tanks more than half-full. You need to be able to evacuate in your own vehicle or some other way, such as patient transport if you have special needs or just a bus arriving at a shelter.

Map Your Evacuation Routes

As part of your evacuation planning, you have to plan for how you get from Point A to Point B. This means that you need maps and written directions to your evacuation points. You also need several alternative routes to get where you're going. There may be traffic, roads may be damaged or closed, or there may be alterations to the services or the traffic flow.

Shut Down, Lock Up, Get Out

As part of your evacuation process, shut off all the utilities. Many houses aren't damaged by the disasters themselves but by damage caused by broken or leaking utilities that flood or burn the houses afterward.

Turning off the power to the house is a matter of flipping the main breaker (or pulling the main fuse block) at the electrical panel. Be very careful when you do this: electricity can kill you even in dry weather. When you shut off the power …

♦ Wear insulated gloves if you have them.

♦ Wear heavy shoes and stand on something dry and nonconductive such as a small wooden stool or a phone book.

♦ Make sure that you're not touching the ground or any object.

♦ Use your right hand (even if you're left-handed). If an electrical shock travels through your left hand and down your body to the ground, it will shock your heart, possibly fatally. Keep your left hand in your pocket so you don't get a shock running across your chest for the same reason.

♦ If there's any danger of shock, such as water on the floor, leave the power on. It's not worth risking your life.

Shutting off the water and gas doesn't require nearly the same level of caution as shutting off the electricity. Both may require the use of a special tool that fits the shut-off valve. (If you need one of these but don't have one, they're usually available for a few dollars from the utility company.)

Prep Facts
You can get advice on how to safely shut off utilities from the power, water, and gas companies. They'll be glad to show you where the controls are and how to safely operate them.

When you leave the house, you need to lock the doors and windows just as you do whenever you're going away for a few days. Make sure that all the latches are shut and that any cheater bars or slide locks are on for windows and sliding glass doors. In addition, lock inside doors that lead to the garage or a sun deck.

Locking up isn't just about securing the doors and windows. Lock up safes or storage units in the house. You also need to lock up your valuables. This can be part of your preplanning: go to the safe-deposit box to drop off jewelry and the originals of documents that you have copies of in your Go-Pak. Some situations require an original document, and you want them to be safe.

Tell People Where You're Going

Whenever you go somewhere, tell people where. (And you need to let them know when you get there.) But you should also have someone outside the region who will act as a message board who can accept a collect call and pass messages. This is even more important if you have a large family or group of people who are evacuating from an area.

Set up designated call times and contact points with friends or relatives outside the disaster area. This way, anyone who needs to know can get information from someone who isn't directly affected by the disaster. This also lets someone outside the region know that if you haven't called by such a time, you may be having a problem and they should alert the authorities.

Follow Directions: No Shortcuts!

When you're evacuating, don't take any shortcuts. Follow the directions that you've already established unless specifically directed to take an alternate route by authorities. Now is not the time to wing it. There is no telling what the road conditions may be like elsewhere. You could end up stuck in a retrograde traffic flow, miles away from your intended destination.

Know Where You're Going

Know where you're going! Have maps of everything: not just your primary routes but of the surrounding areas as well. It's a good idea to have alternate shelters and even alternate destinations, too. For example, if you'd originally intended to go to a shelter northeast of you but the roads are impassable due to traffic, you might want to consider heading due west to a shelter in another state entirely.

Life in the Shelter

If you're evacuating to the home of a friend or family member outside the area, well and good. Have a pleasant visit! But chances are that you're going to be staying in a public shelter. There are some tips for doing well in a shelter that will make your life and the life of everyone around you that much easier.

 Life Preservers _____

For predictable disasters such as hurricanes, many people will secure their houses and then fly out of the area entirely. They'll go off to Six Flags in Texas, Disneyland, or Reno for a week to wait out the hurricane, and then fly home.

What to Bring

A shelter is an aggregate living arrangement. You'll be bringing things out of your house to preserve them, but you don't bring them all into the shelter. Keep anything you don't need in the car or, if you came in a bus or other transportation, in a suitcase. Pillows are good to bring in. Kids need special blankets and favorite stuffed toys. Medical and dietary needs are also good.

What Not to Bring

There are a number of things that you shouldn't bring into the shelter, including:

♦ Valuables

♦ Jewelry

♦ Large sums of cash

♦ Weapons

♦ Anything potentially toxic

Above all, never bring anything illegal to the shelter. This will result in unhappiness for you, including possible expulsion from the shelter and legal ramifications.

Getting Along with People

A shelter isn't a dorm or a domicile. It's a bunch of people who are all feeling tense and worried and who'd rather be home. You're not the only person there and you need to act appropriately:

♦ If it's not yours, don't touch it

♦ Act nice to other people

♦ Wash your hands

♦ Cover your mouth (with your elbow) when you cough or sneeze

♦ Brush your teeth, comb your hair, and bathe or wash daily (when possible)

Dealing with Boredom

After the tension of getting to a shelter and setting up, your biggest job is waiting calmly until you can go home. Although the shelter staff will be doing what they can to keep people interested and occupied, you're going to have a lot of time on your hands.

Make plans to deal with boredom in the shelter. Bring an mp3 player, a computer, or books. Playing cards and games is a great way to pass the time and to socialize within the group. Remember, everyone else is bored and looking for something to do, too. Learning to play bridge or euchre might be just the thing to pass the time.

It's also a good idea to do something physical. Exercise—do tai chi or stretching. If nothing else, you can take a brisk walk around the perimeter of the shelter. Physical activity burns off some of the tension and boredom and generally makes you feel better. You'll sleep better, too. Again, you'll find a lot of people interested in joining you.

If your cell phone or computer can communicate, you may even be able to text message or e-mail people.

Safety in the Shelter

A shelter isn't necessarily a social pressure cooker, but people may be packed shoulder to shoulder in a small, uncomfortable space. Put on your best company manners and make an effort to get along with everyone. If someone else is being cranky, don't take offense; just chalk it up to their own tensions and try to be extra nice to them or give them some space.

Keep your valuables and other temptations to theft out of sight. Don't flash wads of cash and don't leave easy-to-snatch items unattended. You should also keep your children with you, not so much because of the possibility of predators in the shelter, but just because shelters are probably not as childproof as they could be and you'll want to avoid accidents.

Be aware that shelter furnishings—cots, folding chairs, and even tables—may have significant wear. They may have seen many other disasters or just be surplus equipment that's already past its prime. Don't just flop out on furniture. Test it first to make sure it's sturdy.

Shelters usually have cables, cords, and ropes strung around. Although cords may be taped down, they can still present an opportunity to trip or catch yourself. Watch where you're going and tell your children not to run in the shelter. Also be aware that other people aren't necessarily as considerate about claiming an appropriate amount of space.

Time to Go Home

When the disaster's over and it's safe to go home, the shelter will give the all-clear. Gather up your possessions and pack up your car. Check with the shelter about safe routes, possible hazards, and any information they may have about damage or restrictions in the area you're returning to.

Just as when you set out for the shelter, it's a good idea to inform your contacts that you're returning to your home and when you expect to be there.

Returning Home

The sooner you can safely return to your home and business, the more likely you are to have a successful recovery. You have a strong emotional attachment to your home and your business, but the emotional intensity of the situation can trigger an emotional attachment to the shelter as well.

Unlock, Unseal, Turn On, Unpack

When you get home, you should go through a procedure to return to your normal life. You may want to have something similar to an airplane pilot's pre-flight checklist for returning home:

- Unlock the doors and windows.
- Check on your valuables.
- Turn on the electricity (remember the rules for turning off the electricity and take the same precautions when turning it back on).
- Check the house alarms and telephones. Many phones, such as cordless phones, require electricity before they work.

◆ Turn on the gas and check the pilot lights (don't forget the water heater, furnace, stove, even the refrigerator if you have a gas fridge).

◆ With the power on, check the appliances. If they're not working, check the breakers, too. It's very frustrating when the TV or entertainment center isn't working and then you discover that the breaker was just tripped. Also be aware that many appliances have their own built-in circuit breakers, such as microwave ovens, TVs, and hot water heaters. Look in the back of the units for a small pop-out switch or check the manual.

◆ Cable TV and Internet services may require a system reset from the cable company. Check with the company's customer service line.

◆ Even a gas water heater has electrical components, which will have a breaker. Shutting off the gas and then turning it back on may also require you to do a manual reset in addition to relighting the pilot. Again, check the manual or talk to the gas company.

Prep Facts

Some pilot lights require additional heating. It's not enough that you light the stream of gas coming from a nozzle, but the pilot light unit itself must be preheated to allow the gas to flow smoothly. This is by design to prevent gas from flowing if the unit gets blown out somehow. Gas fireplaces may require you to hold a butane lighter under the pilot light to heat things up enough so you can then light the pilot. These units usually also have electronic breakers, which will make things even more complicated.

After you've turned everything on, you need to unpack your suitcases and the boxes of stuff from your car and get it all back into the house. If you've stored jewelry or papers in the bank, you may need to go to the bank as well—but you may also want to leave the stuff in the bank where it is. The reason for safe-deposit boxes, after all, is to store things you're not using regularly in a safe, secure, disaster-proof location. If you don't need the heirloom pearls or the marriage certificate or the stock certificates, let 'em stay in the bank!

Clean Up and Repair

Whenever you have to evacuate, there's a very real chance that you'll come home to a damaged house. The damage could be minor, major, or even enormous. For example, if there was a hurricane, you may have to repair impact damage to siding, roof damage, or worse. Floods will probably require mopping out, drying off, and general cleaning. (Specific types of damage you may have to deal with from many disasters are covered in the chapters in Part 2.)

Check with your insurance company and with the local authorities for information about disaster-recovery resources. Depending on the disaster, there may be local community action as well: you and your neighbors might chip in on a trash dumpster for general clean-up, for example.

As you're making repairs, make note of what didn't make it through this disaster and consider upgrading to something stronger and more durable. Siding that got peeled off by a tornado might be worth upgrading to something stronger, such as concrete Hardiplank-style materials. Window shutters can be reinforced or replaced. Landscaping that might have burned in a wildfire can be replaced with fireproof materials or a rock garden.

Return to Normal Life

The final step to returning home is to return to a normal life. This means that you need to restore the processes and activities that you were engaged in when you evacuated. This may take a while, though. If the school your children attended is temporarily closed while they repair damages, your children will have to go to another school or even stay at home. Some house damage can take a while to fix.

But throughout this process, the most successful survival tactic is to resume your regular activities as soon as practical. This will help you re-bond with your home, your neighbors, and your community. The shorter the disruption in your life, the less emotional impact the disaster will have on you and your family.

The Least You Need to Know

- Every member of your family must have a Go-Pak with them at all times.

- Seventy-two hours of food and water is a minimum level of individual preparation.

- Safe evacuation requires planning and compliance with evacuation instructions from local officials.

- Sheltering in place requires preparation of a secure location in your home or business.

- Life in an evacuation shelter requires the ability to accept the differences in people and to maintain your personal space in the limited space of a cot.

Chapter 6

Preparing Your Business

In This Chapter

- ◆ Why you should make a business-continuity plan
- ◆ How to create an all-hazards business-continuity plan
- ◆ Special considerations for computers and data
- ◆ Planning for the employees
- ◆ Business insurance

Disaster planning isn't just for individuals and families. If anything, it's more essential for businesses because of their greater requirements for careful processes and the need to track production, orders, and accounts. What you may need in your business will vary dramatically from company to company, but all business-continuity planning has some underlying concepts in common.

In this chapter, I show you why you need a business-continuity plan and how to create one, as well as what kinds of things you need to do to deal with your computers and data. I also have some suggestions on business-continuity planning for your employees and business insurance.

Business Continuity and Beyond

According to the National Academies of Science, fewer than 4 percent of U.S. hospitals are prepared to handle any type of disaster scenario, from a hurricane to a terrorist bombing to an industrial accident. Additionally, few companies even run disaster-planning drills on a regular basis. So when something happens that affects a company, such as a chemical spill, a flood, or a pandemic flu outbreak, everyone from the employees to the local hospital staff to the community at large feels helpless and out of control. That's when things fall apart.

Disasters are not just possibilities. A disaster will happen in every community, whether it's another Katrina or Chernobyl. Consider this: a few years ago, 1.2 million cubic feet of propane was accidentally vented across U.S. Hwy 1 out of the Florida Keys, causing the road to completely close. This is the only road leading in or out of the area. Local emergency services didn't know how they were going to stop the leak. The company that was at fault was relying on local services to deal with it. Because no one had drilled for such a disaster, no one knew how to respond. Something that could have been a short-term emergency turned into an all-out disaster.

A disaster means business disruptions, loss of income, and a potential loss of customers, no matter what sort of business you run or work in. You need a *business-continuity plan* to tell you what to do in the case of an emergency. A business-continuity plan is essential for small businesses that don't have a lot of resources to meet disasters with, and for larger businesses, which have more to lose. And a business-continuity plan is essential for very large businesses and government organizations, which are frequently inflexible when it comes to having business or procedure disruptions.

A solid, well-tested plan minimizes the disruption to your business's daily operations, as well as the disruptions to your employees and customers. It's essential that your business have a plan that identifies the potential emergencies the business may face, information on how to prevent or at least mitigate the risk of emergencies, and procedures on what to do if a disaster strikes. This plan must be communicated to the employees and periodically tested and re-evaluated.

def•i•ni•tion

A **business-continuity plan** is a comprehensive set of procedures that tells you how to respond to disasters. It includes such things as protecting the company's assets, data, and people; mitigating the damages; and arranging for temporary business facilities elsewhere, as well as who to contact for emergency authorizations and how to restore business operations.

Getting Ready for Disaster

Businesses need to ensure that they have a plan and that they drill the plan regularly. Whether you work for a large, multi-national company or a small, privately held firm, use the following guidelines to be ready when disaster strikes.

First, you need to create your own plan. Many companies have purchased disaster plans and never taken them out of the shrink wrap. They simply bought a generic disaster plan to get an insurance discount and never opened the book and read it, let alone used it. If they actually attempted to use the plan, they'd quickly learn that the plan is not an all-hazards plan. Rather, it's disaster-specific. So if they have three types of problems occurring simultaneously, they'd be flipping back and forth in the book trying to figure out what to do. That's when they'd discover that certain sections of the book contradict each other when used at the same time.

So rather than purchase a pre-written plan, create your own. Most industries have mandates for a safety officer—someone who is supposed to be registered and educated in industrial safety. Make sure you hire someone to fill this role and that this person writes a detailed hazard plan for your company. Again, this plan should be an all-hazards plan that covers a series of cascading events.

Next, you need to think of the big picture and ask what could possibly go wrong. When you create your all-hazards plan, think in terms of everything that could possibly go wrong. Let's suppose you're writing a disaster plan for a chemical plant. The reality is that if the chemical plant blows up, whether due to a terrorist bomb or employee error, that explosion is going to cause catastrophic events that will have a chain reaction.

The explosion not only causes chemicals to spill into the community, but it also causes power lines to fall to the ground and raging fires to start in nearby businesses. With the spilled chemicals come water contamination, and the downed power lines put people at great risk of electrocution. The raging fire at the neighboring manufacturing plant releases toxic fumes into the air. Then to top it off, it starts to rain … a lot. Now you're not only dealing with a chemical spill, but also fire, water, and ground contamination; electrocution; toxic fumes; and flooding.

That's why your disaster plan must address the entire disaster, not just the plant explosion. It's prudent to bring in outside consultants to help fine-tune the plan. The market is full of external organizations that help companies create and drill plans specific to their industry. Just because you have an all-hazards plan doesn't mean you're ready for all hazards. You need the insight of an external organization that can help you see your plan and your impending disaster in a new light. Without this outside perspective, you could very easily be creating your plan in a vacuum, overlooking key elements that would save your company money, time, and even staff.

After the plan's in place, drill the plan twice a year. All businesses, from small, family-owned firms to major corporations, need to accept that conducting disaster drills at least twice a year is a normal operating expense that cannot be ignored. Set aside a few days each year for your employees to run the drill, and pay your staff their usual wages during this time. For some companies, this may mean ceasing operations for the day so all employees can be involved and do their part.

Life Preservers

Many companies make the mistake of having a plan that they've never tried. When disaster strikes, they discover that it doesn't cover what the business actually needs to do. Always consider an untested business-continuity plan no better than having no plan at all.

The drill should cover more than one type of disaster scenario so you get a cascade of events. These drills must be complex and involve all aspects of your plan. If your plan does not break down during the drill, then you have not drilled hard enough. You need to take the plan to the point where you can identify every weakness. Only then will you have the true picture of what your organization can handle and how to compensate for anything lacking. If companies fail to take the drill to this extreme level, then we're going to see a lot more Hurricane Katrina scenarios—where demand exceeds resources—as our population becomes larger and our world becomes more complex.

Finally, you need to get the entire community involved. When you drill your plan, you must bring in any outside community help that your plan calls for, such as medical staff, fire rescue, EMS, police bomb squads, and so on. You simply cannot run any drill in a bubble with actors portraying the needed roles. Running a drill means going out into your local area, coordinating the drill with other organizations, and initiating action within your community.

Creating an All-Hazards Business-Continuity Plan

The things you need to plan for in an all-hazards business-continuity plan are different for each business, but the general steps for creating an all-hazards business-continuity plan are the same:

1. If your business already has a business-continuity plan, review it. It's likely to have many elements that are still useful.

2. Identify the types and impacts of emergencies. There are many different types of emergencies that you need to plan for as part of your business-continuity plan, such as the following:

   ◆ Environmental disasters: floods, fires, hurricanes, earthquakes, blizzards, tornadoes/windstorms, volcanoes

   ◆ Other disasters: bomb threats; strikes; employee sabotage; terrorist attacks; riots; building fires; gas leaks; departure, illness, or death of key employees; supply-chain disruptions

   ◆ Utility or service interruptions: water, power, heat, phones, Internet

♦ Equipment failures

♦ Data theft and security failures

♦ IT failures

As part of the assessment process, you should also try to identify the likelihood of these emergencies. For example, it's extremely unlikely that an earthquake will affect a business in North Dakota, but the chances of a blizzard disrupting some aspect of the same business may be an annual event.

3. Assess the financial and organizational impacts of the various business risks. Each of the emergencies identified in the previous step will have a specific effect on the business. Tabulate the types of impacts, including the potential for interrupting standard business operations.

4. Assess the existing emergency procedures. Identify the existing procedures for emergency situations and the people responsible for emergency procedures. Also build a list of emergency contacts, emergency services, and people in charge during an emergency.

5. Prepare for possible emergencies. All emergencies require the same basic preparation for maintaining operations and preserving business essentials, in particular:

♦ Designing back-up and recovery strategies

♦ Identifying essential documents and records

♦ Arranging for a continuing off-site storage system for back-ups and essential information

♦ Arranging for an alternative business site

♦ Developing plans for administration, operations, financial, sales, and customer support continuity

♦ Reviewing the business's insurance coverage

♦ Setting up emergency financial and operations authorization procedures

6. Plan the disaster recovery. Identify the disaster recovery team(s) and create the procedures they will follow to mitigate and recover from the effects of the disaster.

7. Plan the business recovery. Identify the business-recovery team and create the procedures for bringing the business back online. Prioritize the key business services and functions and plan to focus the business's efforts accordingly. Consider the potential need for a media relations effort.

8. Test your business-continuity plan. Plan the tests to simulate a real emergency as well as possible. (Consider having only a small group know that the test is really a test.) Identify the testing team, the test parameters, and which emergencies you're going to test for.

9. Revise the plan as necessary based on the test results, then retest. Don't shortcut the process: if an emergency procedure fails to meet your requirements during testing, it won't work in a real emergency when you need it to. You need to have procedures that you know will work for you.

10. Provide general employee training. After the business-continuity plan has been approved, create employee training materials, handbooks, emergency procedure sheets, and so on. Train everyone in their appropriate emergency functions.

11. Update the business-continuity plan regularly. Re-evaluate your plan every 6 to 12 months to make sure it still fits your needs. Make sure that the people named for the various teams and task units are still available and that the contact numbers and information are correct. Provide an employee retraining session, both to refresh peoples' memories and to provide training for new employees as well.

Prep Facts

Be really wide-ranging in the kinds of disaster scenarios you create. Keep asking "What if?" and see what kind of problems you can imagine. Also come up with combinations of problems, such as a flood and the loss of key people or a pandemic and a strike and an earthquake. Many great disaster stories are not the result of a single crisis but of multiple serious problems striking all at once.

IT Considerations

Some companies have always made disaster plans. Usually, these were companies in high-risk areas: Southern California, the Midwest, and the Gulf states. Being hit regularly with disasters has made many businesses in these areas mindful of the need for planning. Ironically, only in the last 20 years has there been a strong move in American business for disaster and continuity planning, and it's been computerization that has done it.

Apart from the direct value of the physical computers and the software, the data stored on computers can be invaluable. As part of any business-continuity plan, you must make provisions for the computers.

Depending on the level of computerization at your company, you may need to have a separate IT continuity plan that is part of the overall business-continuity plan. If you're a small company, data backup and system recovery may be nothing more than picking up a few computers at a local dealer and loading accounting and office productivity software and restoring data from your latest backup. On the other hand, large companies can have entire buildings of computer servers and hundreds or thousands of individual workstations. The disaster planning for such an organization requires a comprehensive and thorough examination of the goals, costs, and timetables for disaster recovery and resumption of normal operations:

1. Identify the critical computer systems in your company.

2. Develop a back-up system.

3. Arrange for a "hot-site."

Examples of systems that are likely to be critical to your company are servers, accounting, e-mail, and factory automation or production monitoring software. Many companies also have the phones running as part of the computer network as well. The best method is to list all the computer systems currently being used and assign a rating for their criticality to the company's daily operations and a priority for restoring them in a disaster.

Many companies require daily backups of critical information from individual computers to a server array, which is in turn backed up nightly with high-speed tape systems. These backups may simply be mandated by policy but there are many ways of creating an automated backup, where a server on the company's network taps the specified individual computers for data and basically sucks it out of the hard drive to the server. Most companies store back-up tapes in a fireproof vault either on-site or off-site, minimizing the potential for lost data.

If your company burns to the ground, you're going to need to restore your data on a similar computer system with a similar network, or at least a skeleton system that will support your essential functions while you make other arrangements. Hot-sites are locations you can use to restore data and operate in an emergency. Some companies have hot-site swap agreements with other companies who have similar setups, so that if one or the other can't use their normal facilities, they can double up on the other company's computers.

 Life Preservers

If you can't arrange to swap hot-sites with anyone, there are services that provide mobile hot-sites built into vans or even semi-trailers, which can drive to a building site or an area and help the company restore its data operations. These aren't cheap, but for larger companies with a lot of computing requirements, they can be the perfect solution.

Plan for the Employees, Too

After years of education and budget planning, a company finally realizes that instituting a business-continuity plan, ensuring data redundancy, and even having an alternative operations site, are prudent investments for the future. Anticipating employee needs, the company arranged for fuel, food, and water at the alternative operations site. No sooner did the company announce that they were fully "disaster-ready" than a natural disaster struck. In a matter of hours, the community suffered devastating losses as homes were destroyed, power and water services were cut off, and commercial buildings were rendered uninhabitable. The plan itself worked perfectly—but the employees didn't show up. The otherwise responsible and loyal employees deviated from their expected behavior.

It's not enough to ensure business continuity, you must ensure personnel continuity. Most employers count on one of four factors to ensure their employees report to work: pay, appreciation, loyalty, and fear. Even with above-average pay, employees and even management will secure their home and family before reporting to work. When asked, these individuals cite personal values, family priorities, and "seeing what is important" as the reasons for their change in behavior, yet after the disaster is past, most of these individuals return to their pre-disaster work ethics and schedules.

As part of business-continuity planning, businesses must maintain good employee relationships and find ways to encourage their employees to show up. Strategies for doing this are discussed in Chapter 15.

Insurance, Assurance, Peace of Mind

Adequate insurance coverage is a vital part of a business-continuity plan. A business needs to have enough insurance and the right kinds of insurance so that they're not going to pay out of pocket for basic expenses. When you have identified the various risks and their likelihood as part of the planning process, check with the business's insurance carrier to see if you have adequate coverage for the identified risks.

The Least You Need to Know

- ◆ Business-continuity planning includes planning for your physical systems and data, the business location, and your employees.

- ◆ Identify and institutionalize critical business processes and relationships to facilitate continuity of operations.

- ◆ Understand the factors that influence whether employees and customers show up after a disaster.

- ◆ Educate your staff about their role in the continuity of operations plan and how they can participate if personal demands prevent on-site attendance. Giving options helps employees make informed decisions that often favor the business in both the short and long term.

- ◆ Ensure that your business maintains key relationships with employees and customers in good times and bad.

Part 2

Detailing the Disasters

Even though all disasters disrupt life and your normal flow, there are differences in how to deal with the specific types of disasters. This part talks about the details for eight different disasters. Some of these are more common than others, but it's good to know how to deal with all of them because they all could happen. There's information in each chapter about the disaster itself and how it happens, what kind of warning you'll get, if you should evacuate and when, how best to survive the disaster, and related hazards.

Chapter 7

Floods

In This Chapter

- ◆ When and where floods are likely to happen
- ◆ General flood preparations
- ◆ Tracking floods and evacuating
- ◆ Flood hazards to watch out for
- ◆ Drying out afterward

More than 90 percent of the world's population lives within 10 miles of a major body of water and thus is subject to flooding. Seasonal floods and rising waters have been a part of life since ancient Egypt and before. Many cultures have evolved to adapt to the flooding typical of their environs. Flooding, however, is not always predictable or a cultural norm. These are the times when flooding becomes a disaster.

In this chapter, I tell you what a flood is and the various causes of flooding, when and where floods are likely to happen, how to prepare for a flood, and how to keep an eye on potential flooding conditions. I also tell you about flood hazards you may not be aware of and how to clean up and dry out afterward.

What's a Flood?

The definition of a flood is incredibly important. All floods involve too much water in the wrong place at the wrong time, but there are different sources and accumulations of water that can make a big difference to what you do and how you recover from it.

A flood is any water in contact with the ground that rises above the normal level or grade. What's important about this definition is that it meets the legal definition of a flood for insurance and disaster relief. If you don't have a flood rider on your insurance policy and you're outside of a flood zone, then you have no federal disaster insurance, either.

Groundwater

Rainwater that falls and collects is not a flood. On the other hand, groundwater that rises is a flood, even though it may be raining on the mountains and the water comes down underground and it comes out of the base of the mountain.

Surface Water

Surface water flooding is caused by a rising river or a confluence of the rainwater, which then gets together and moves. Where it pools is the flood. If your front yard is covered in standing water, it's standing water and it's rain, but it's not a flood. However, if my house is higher than yours and my front yard collects standing water until a retaining wall bursts and it floods downhill into your house, then that is a flood.

The point of all this is that all floods involve lots of unwanted water, but lots of unwanted water may not technically be a flood.

Surges and Waves

Floods are frequently caused by surges. A surge is where a large amount of water is added to a river or other body of water, which causes a swell of water to move outward from the entry point. A river could be suddenly hit with a huge amount of water from a mountain after an unusually large spring runoff, or a second river that feeds the first could have gotten an unusual amount of water miles upstream from heavy rains.

This can even have a cascading effect: the levees along a river that's near flood stage can be holding just fine even though the river is much higher than normal. But if one levee upstream breaks and a surge of water hits the next levee, that additional pressure can be enough to burst that levee, too, and it can go for miles downstream. Even if the levees hold, the water may be high enough to crest over the levees.

Storm surges can also happen as a result of high winds from hurricanes, particularly near coastlines. In this case, it's very much like surface water you push around with your hand. The wind moves one section of water upstream pushing it higher and higher in the hurricane's front, moving a wall of water higher than the surrounding ocean. As long as the wind is blowing, the wall keeps rising, and then when the wind dies back, the waters recede. Storm surges can be as brief as a few minutes or, as happened in 2008 with Hurricane Ike, can last as long as 12 hours.

Surging water doesn't just happen with hurricanes; it can go in front of *any* wind storm. In Florida, there are surge floods along Lake Okeechobee. Storm surges can also be very destructive when they're pushed up a waterway. Pushing an additional 4 to 6 inches of water up a river that's already close to flooding can have disastrous effects. In addition, because the mouth of the river is probably much wider than the channel, the water may be pushed even higher than the storm surge itself, causing the river to overflow suddenly and dramatically. But as the winds drop, the waters flow back out, dragging tons of debris back downstream and hitting already damaged structures with tons of water from the other side.

Surges aren't waves. Tides cause waves, which are just an energy front moving along a body of water. The level of the water goes up and down with the wave, but the individual molecules of water don't move forward or backward. In a surge, however, the water is being moved by the winds of a storm.

When and Where the Waters Rise

For a flood to happen, you have to move or add water. Floods happen in some very specific times and places, depending on where you are and the reason for the floods. The type of flood you may experience can differ dramatically as well.

Flood Seasons

Floods can happen at any time, depending on the type and location. One of the most common types of flooding, river flooding, is the result of adding water. For example, during fall and winter, the Yangtze River is safe right to the very edge of the river banks; but in spring, the water volume can double from snowmelt and rains.

Other types of river flooding, such as the floods that happen in the Midwest, occur more often in spring and summer as a result of heavy rains and winds. Some surge flooding is seasonal because it's associated with other seasonal weather conditions such as hurricanes, monsoons, and so on.

Flood Reasons

A river flood happens when the river overflows its banks, so the entire river is bigger than usual. The water all goes over at the other end. This can happen for any number of reasons, including heavy rainfall, excessive snowmelt from warm weather or a very heavy snowpack, or storm surges.

As mentioned earlier, floods can also happen at the foot of mountains when water that falls on the mountain soaks into the ground and then comes out near the base of the mountain. This kind of flood is very unpredictable because there's little visible sign that it's going to happen.

The first drops of rain during a wildfire may be greeted with joy, but too much rain can cause flooding. Wildfires can scorch the ground and make it difficult for water to soak in. As a result, rain that would normally have soaked into the ground can pool and run downhill, causing flash floods and mudslides.

Up until recently, you wouldn't have a hurricane in the Midwest. Now, you can: in recent years, hurricanes have moved farther west and north than in years before.

Flash floods can result from large amounts of runoff from heavy storms or from sudden surges of water, such as levees or ice jams breaking. They frequently happen in mountainous or steep terrain, or where water doesn't soak into the ground, such as in deserts and after wildfires. The biggest danger of flash floods is how sudden they are. Flash floods can cause "walls of water" 6 feet high and higher in just a few minutes; heights of 10 to 15 feet in canyons are not uncommon. If you live in an area where a flash flood is likely, be ready to evacuate at literally a moment's notice. That may be all the time you have.

Prep Facts

One very unusual type of flood, a *lahar,* is actually volcanic in origin. A lahar happens when snow or glaciers are melted during a volcanic eruption. A lahar can also be caused by a glacier melting at the top (usually from volcanic heat below), forming a lake near the top of a mountain with a glacial ice wall holding the lake in place on the downhill side. When the water breaks free, it pours down the mountain, picking up debris and mud as it goes. Lahars can easily move at 60 m.p.h. or more for miles. Major portions of the South Puget Sound area in Washington State have been covered with yards of mud from lahars from Mt. Rainier as recently as 500 years ago. The Osceola lahar 5,600 years ago buried 130 square miles under 450 feet of mud. The Puget Sound has a warning system in place to alert residents of an impending lahar.

Flood Places and Paces

Floods can happen pretty much anywhere:

- Near rivers
- Near coastlines
- Near lakes
- Near mountains
- Anywhere there are wildfires
- Anywhere it can rain very heavily

In addition to the reasons for floods, you have to also think about how fast the flood might be. For the most part, where you are determines how fast you're likely to be flooded. If you're below the level of a river, it'll flood quickly: the water just has to crest whatever's keeping it away from you, and it's all downhill from there. On the other hand, if you're in Flood Zone X (an insurance designation for the least likely flood areas), you have little risk of groundwater and surface flooding.

Pre-Season Preparations in Detail

Preparing for a flood doesn't mean you need an ark in your backyard, but you do have to make some significant preparations to your home and your community.

Build Up and Build High

The French knew this in New Orleans: it's why the French Quarter was built on high ground. When the levees break, the French Quarter is largely untouched because it was engineered to be a really slow flood. The water may rise enough to cause damage, but the flood is slow enough that you have enough time to move furniture, rugs, and possessions upstairs out of the way. Many other places in the Midwest are similarly situated: the town's founders knew from experience that they'd get flooded eventually; the trick was to build houses so that the floods did the least amount of damage as slowly as possible. You can see this principle elsewhere, too: many ancient Egyptian buildings on the Nile (which flooded annually) were built on high ground or away from the fastest waters.

Levees and Dikes

Levees and dikes are probably the oldest means of flood control: you know the river bank's going to overflow, so you make it taller. And because you know that the flow of water is going to erode the stuff you put there, you can plant erosion-control stuff in the dikes to make them last longer.

What's the difference between a levee and a dike? Not a lot, really. Both of them are embankments that raise the sides of a river and prevent higher-than-normal waters from flowing over the sides and causing flooding. The Anglo-Saxon root word for dike is also the same as that for ditch, so a dike was originally either the embankment or a ditch next to it. Nowadays, they're the same thing.

A number of cities have used dikes to keep water from flooding them. The city of Amsterdam is about 6 feet below sea level. The dikes are the only thing that keep the city from flooding. Half of New Orleans is at or below sea level, making dikes essential there.

Ironically, a good deal of the work that the Army Corps of Engineers has done to straighten the Mississippi River and make it easier for river traffic has increased the danger of flooding. The river is now faster, giving the surrounding riverbed less time to soak up excess water, which has increased the risk of damage from storm surges.

Levees and dikes were originally made of earth and wood. A hollow wall of planks was built up, reinforced with poles to hold the planks in place, and then filled with earth and rocks. The problem with wooden embankments is that they rot fairly quickly and can wash away, so they're not used anymore. Today, levees and dikes are manufactured. They're usually concrete, masonry, or steel. One type of manufactured embankments is interlocking concrete blocks that are then filled with sand. (This is something similar to an earthen levee in reverse, with an exoskeleton.)

In other countries (though not so much in the United States), levees are made of steel. They're frequently involved in a lock system, which is distinctly different from reinforcing walls. The locks prevent wave action from causing erosion, so the retaining walls also keep the river banks in shape and keep you from losing real estate.

Levees and dikes can fail in three ways. They can be breached, where they simply break, collapse, or wash away. This can be caused by impact from debris in the water or just from the pressure of the water on the embankments. Levees can also be overcapped, where water comes over the top (such as sloshing water over the edge of a bathtub). They can also be undermined, where water erodes the foundation of the embankments and makes a channel underneath (such as a dog burrowing under a fence). Capping a dike with rocks or concrete can help keep it from eroding. The rocks help push the levee down and push it into the slot to prevent a channel from forming.

Sandbags

The last type of embankment prevention is the sandbag, which is nothing more than a portable dike. Sand by itself isn't very structural and it washes away very quickly—picture a sandcastle after being hit by a few waves—so the burlap sacks holding the sand keep it together. The sandbags also make the sand easier to move and stack. You stack the sandbags, put sand behind them and sand in between, and poof! you have a quick earthen levee. The water can go under or over, but when it tries to go in between, the water moves the sand and plugs its own hole.

Sandbags are very convenient and relatively quick and easy to create. They just take a lot of muscle to shovel wet sand into bags. Stacked sandbags aren't structurally finicky, either. You can stack them up without having to use a level (unlike creating a concrete block wall). They also can be used as reinforcement for buildings or even more permanent embankments that may be in danger of getting knocked over by water pressure.

Waterproofing

If it's likely to flood in your area, you want your house to be an island of dry ground that's well above the waters. Wherever possible, start with the highest piece of ground you can and add as much height as possible. This will work for surge, wave, and surface water.

When you're waterproofing for a flood, you're trying to keep groundwater from soaking up the concrete slab or the basement. You can use things such as elastometric paint, waterproof concrete, and so on. It may help you realize you're building an inside-out swimming pool, where all the dry stuff is in the pool and all the wet stuff is outside. You also need to set things up so that you don't waterproof everything and then find that the surface water has poured into your basement through a window.

Life Preservers

When you're buying or building a house, find out what you can about the past flood levels and how they affected the house. In extreme flood conditions, you might want to build a house with a very high crawl space or one on stilts.

First Warning and Tracking the Flood

When it comes to following warnings and tracking a flood, you're back to the reasons it's flooding, which almost always depend on the season.

For example, in Arizona, it floods in the rainy season. The desert soil doesn't absorb much water at all, so if it's raining, it's flooding somewhere. On the other hand, if you're somewhere in the Rockies and you see the first piece of ice melting off the trees, then you know the snow cap is about to melt, which means that a flood is not a matter of if, but when. And if there's a hurricane on the radar, there's going to be a storm surge somewhere.

The next thing to determine is how big the flood's going to be. River levels have been predicted for millennia by measuring the rise at a given reference point upstream. Storm surges in the ocean can be measured with coastal buoys monitored by satellites. This gives a very accurate picture of the ocean's water level.

Most floods are caused one way or another by storms: too much rain, too much wind, or both. Heavy rains in the Northern Midwest almost always mean flooding somewhere, as does an unseasonably hot day in early spring. One of the best ways to keep an eye on the potential for a flood is to watch the bad weather. Many floods have been caused by a sudden downpour of several inches of water.

Final Warning and Evacuation Issues

If there's likely to be a flood in the area, you can count on the news to keep you posted. If the river's nearing flood stage, there are going to be regular bulletins.

Get Out of the Way

The big issue when you're evacuating from a flood is that there is really nothing you can do about it. There's going to be a flood. You may be able to mitigate your damages with sandbags, but you ultimately *have* to get out. The sooner you do, the more you can take with you.

When you evacuate, you need to be sure you're heading away from the flood. You'll probably know the direction the flood waters are coming from, but it may be difficult finding an escape route that isn't flooded. Heavy rains may have washed out roads or blocked access to an otherwise effective route with a mudslide.

If you're driving and the road you're taking is flooded, even if it doesn't look like much water, don't take it. Your car can be washed away by a surprisingly small amount of water. Almost half of all deaths due to flooding are related to vehicles that are swept away. Also, if the flood waters are rising and the car you're driving stalls, abandon it and head for higher ground.

Grab and Go

You should have been keeping an eye on the news if there was a visible flood hazard. However, even if you have been watching the level of the nearby river, a flood can occur very abruptly due to the failure of a levee upstream. In this case, you may have only a few minutes to evacuate. If you get the word that a flood is coming, grab your Go-Paks and leave—*now!* You have no way of knowing how big, how fast, or how full of debris or pollution the flood waters may be. A flood isn't something to trifle with.

If you simply can't evacuate due to a lack of time or no escape route, grab your Go-Paks and head for the second floor of your house. Keep the access to your attic available (if you have one) and see if you can get to the roof if you need to get higher still. You may be stuck up there for a while depending on the conditions, so bring some heavy outerwear. You'll also need to bring food and water. (Don't even think about drinking the flood water; it can be full of sewage or chemicals.)

Life Preservers

If you have a second floor and it looks like you're likely to get hit with flood waters, move as much of your furniture and possessions upstairs as you can. This way, the flood won't soak them. It's also a good idea to wrap things that are susceptible to moisture damage, such as books, in plastic garbage bags.

When the Flood Hits

Your location, the terrain, and the causes of the flood will all affect how the flood is going to hit. The one rule of thumb for survival is that flood waters will always rise faster and spread quicker than you think. The danger signs of imminent flooding are:

- Rapidly rising water in your area
- Runoff water turning muddy (indicating that debris is being swept downstream by an excess of water)
- Water approaching you
- A roaring sound

If you see any signs of imminent flooding, *get out immediately.* If you can't safely drive away from the rising water, head for the roof or high ground.

Related Hazards

Floods come with a lot of related hazards, most of which are invisible until it's too late.

Swift Waters

Almost all floodwater will be murky and you won't be able to see anything in it or through it. Any moving water is extremely hazardous, even if it may not appear to be moving quickly. An inch of water moving at 10 m.p.h. will sweep a full-grown man off his feet. Four inches of water moving at 8 m.p.h. can sweep a car down the road. If you get caught in 4 inches of water and you aren't able to get up, you're likely to drown.

If the water is moving at 10 m.p.h., there will be eddies and whitecaps. A stick thrown into 10 m.p.h. water will be 100 yards away in four seconds. And even if the water appears to be calm, there may be currents below the surface that you can't see.

One last thing: you're not used to predicting where water is going. A stream of water may look like you can tell where you're going, but there are currents, turbulence, and things crashing into you that make it impossible to guess. In addition, simple hazards that you wouldn't think twice about under normal circumstances can be dangerous during a flood. A pothole 2 inches deep can catch your foot enough for the current to knock you over and possibly sprain your ankle or worse.

Floating Hazards

As if swift-running water isn't bad enough, you have to also deal with floating debris. Floods pick up whatever they find and carry it with them until it catches on something. What's floating in the water could be literally anything and none of it is anything you want to connect with.

For example, suppose that you've determined that the water in front of your house is only a couple inches deep and is moving at no more than 2 m.p.h., tops, so you decide to slog across the street to see if your neighbors are okay. Here are some of the things that could intersect your ankles at 2 m.p.h.:

- Six feet of rope
- A 2×4 hitting you end-on
- A piece of roof sheeting with roofing nails sticking out
- Tree branches
- Fishing line with fish hooks

Desert areas frequently have arroyos or washes, dry creek beds that fill with water during the rainy season. People who live near them learn to avoid swimming or getting near the arroyos immediately after rains because rattlesnakes are frequently swept downstream. Rattlesnakes are normally cautious and will avoid contact with people as a rule, but they don't swim well and can't maneuver in water. If they get near you in the water, they will attack. This is doubly true for snakes such as water moccasins, which aren't afraid of people to begin with (and are much deadlier).

But you don't have to meet something as nasty as a rattlesnake. A cat that's been swept downstream will be terrified and will attach itself to anything it can climb. You can end up with deep gouges and bites all over your face and head as the cat does anything to stay as high above the water as possible. Even a sewer rat will run up your pants leg seeking safety for the same reasons.

Storm Warnings

Flood debris can hide life-threatening conditions. For example, manhole covers frequently are popped off by waters rising out of the storm drains. A 2 foot × 2 foot piece of cardboard can completely obscure an open manhole. If you fall down the manhole, you're likely to die, either from the impact or because you'll be swept away and drowned.

Shocking Stuff

Water and electricity do not mix. GFI (ground-fault interrupter) circuits were invented because a regular breaker won't work when the circuit is grounded by water. Most housing codes require the use of GFI equipment anywhere that's likely to be near water in your house, such as bathrooms and kitchens, but GFI equipment is much more expensive than plain light switches and outlets. If you're in your living room standing on a wet carpet, you can flip a light switch on the wall, electricity can arc from the switch through your body and out your feet, and biff! you're dead.

Whenever you're standing on something wet, you have to assume that you're grounded and at risk for electrocution. Think of it as if you're in the bathtub: don't touch anything that's plugged in, don't touch light switches, don't get near power outlets, and never turn anything on. (Don't forget that even a gas oven is plugged in for the clock, timer, and thermostat.) If you're in a house that's flooded and the power is on, turn off the power at the main panel, but only if it's safe to do so. If you can't stand on something dry and flip the main breaker switch, don't do it.

Outside, you should be aware of possible downed power lines. Even a power line that's landed 20 feet away or more can deliver a serious or fatal jolt of electricity if you're standing in water. Also look out for unexpected sources of electricity that could be hazardous. For example, if your neighbor has a decorative electric light pole, it may have shorted out from the water on the ground. Leaning up against it could be all you need to get a nasty shock.

Bad Waters

Floating hazards are complicated by the presence of sewage, chemicals, and other toxins. Sewage is a given: if the street and ground are flooded, sewage will have been washed out of the sewers and contaminated the flood waters. Other dangerous chemicals can include pesticides, herbicides, petroleum products, or poisons.

> **Prep Facts**
>
> If you get a scratch, cut, or puncture wound during a flood that comes into contact with flood waters, be sure to get a tetanus shot as soon as possible.

For example, if you're downstream from a garden center, the water can pick up a concentrated dose of insecticides and chemical fertilizers. Even a couple bottles of malathion or bags of casoron granules in your neighbor's gardening shed can add a potentially dangerous concentration of chemicals to the water in front of your house.

Post Flood Issues

The flood will eventually dissipate. You're now going to have to return home and clean up everything.

Drying Out

The first task is to dry everything out. If it can be dried and you can get it to dry quickly before it molds, mildews, or fills with bacteria, you're okay. Look for drying agents, forced air dryers, and so on.

You also have to decide what isn't salvageable and get it out of there as fast as you can so it doesn't become a moisture and mold problem. If you don't have any alternatives, shovel things into a wheelbarrow and pile them on the front lawn. It'd be a good time to team up with your neighbors to rent a dumpster so you can all get rid of the unsalvageable stuff. (Throwing damaged personal items into a dumpster where you can't see them may help you deal with the pain of losing treasured possessions.)

You may think that you can dry out the drywall and the insulation enough to save them, but don't do it. Drywall or insulation that's gotten soaked is going to trap enough moisture to develop mold before it dries out. The house will then be full of mold and is likely to develop dry rot and staining from mold spores on the walls and ceilings. Rip out every piece of affected drywall and insulation and throw it away. It'll be cheap enough to replace.

> **Storm Warnings**
>
> Always use gloves whenever you're handling materials soaked by flood waters to avoid possible infection from sewage or other contaminants.

Cleaning Up

Some things can be cleaned while wet. For example, you can throw clothes, towels, and linens into the wash when they're wet. (Even for coloreds, add a tablespoon of liquid bleach to sterilize them and wash them in the hottest water possible that won't damage them.) For other items, such as area rugs, you need to clean off as much of the mud as possible, wait until they're dry—big drying fans will be helpful—and then vacuum up the dried mud and clean them again.

Clearing Out

You never know what's in the water. It could be chemicals, oils, almost certainly sewage and greywater. Plan on spraying everything with a mild bleach spray (1 tablespoon of liquid bleach per quart of water) to get rid of bacteria. You can't sterilize something that's still muddy, though; you have to remove the mud to get to the thing you want to clean.

Speaking of clearing out, the local water company is going to need to test your drinking water for E. coli, cryptosporidium, and a variety of other unpleasant waterborne bacteria, and then flush the pipes before they'll certify the water is safe to drink. If you're on a well, check with your local water authority or a water engineer before you use the water for anything but external use.

You should discard any food, including fruits and vegetables, directly touched by the flood waters. You can safely eat canned or packaged food after you have washed off the containers and sprayed them with bleach solution. You must also wash and sterilize cooking and eating utensils carefully.

The Least You Need to Know

- Floods can happen anywhere at almost any time.
- Even shallow flood waters can be deadly.
- If flood waters are approaching, evacuate or head for high ground immediately.
- It only takes 1 inch of water to sweep a man off his feet.
- It only takes 4 inches of water to wash a car away.
- A flood can still be deadly after the water recedes. Beware of contamination.

Chapter 8

Hurricanes

In This Chapter

- How hurricanes form
- Getting your house ready
- Hurricane watches and warnings
- Evacuating versus staying put
- Coming home

Hurricanes are a little different from most other disasters because you can see them coming. They even have a season, so you know when to expect them. Hurricanes typically develop slowly and move slowly and somewhat predictably, providing plenty of warning to those in harm's way.

The disadvantage is that hurricanes do an awful lot of damage, so they require a lot of preparation.

What Are Hurricanes?

Hurricanes go by several names: tropical cyclone, anti-cyclone, even typhoon. But it doesn't matter: they're all the same wind that blows nobody good.

def•i•ni•tion _____

> A **hurricane** is a huge weather front up to 600 miles across with spiral-
> ing winds between 75 and 200 m.p.h. They can last for a couple weeks
> from their birth as low-level thunderstorms to when they finally peter out.
> A tornado, in contrast, is a fairly local occurrence that pops up suddenly
> and lasts for an hour or two and doesn't usually travel more than a couple
> dozen miles.

Hurricanes happen when a cluster of thunderstorms drifts over ocean
water with a temperature of 80°F or higher. Trade winds blowing in
opposite directions squeeze the warm air rising from the surface of the
ocean and start it spinning. If the conditions are right, the warm air
rises to about 10 miles above the ocean's surface, where it cools off and
drops back down. This creates a low-pressure area on the ocean's surface, drawing still more warm air up, and voilà! The storm system is
self-feeding, but it isn't a hurricane yet, not by a long shot. Potential
hurricanes start out as tropical disturbances, which are just thunderstorms with a light cyclonic spin. If the wind speeds pick up to 23 to
39 m.p.h., the thunderstorm becomes a tropical depression. From 40 to
73 m.p.h., it's classed as a tropical storm, and above that, it's officially a
hurricane.

Hurricanes come in five categories, depending on how fast the winds
are moving. Category 1 winds are 74 to 95 m.p.h. and Category 5 winds
are more than 155 m.p.h. You don't usually have to evacuate until you
hit Category 3. But just because you may be dealing with a "little" hurricane, local collateral damage from storm surges, flooding, and tornadoes
can be just as destructive as a Category 5 Katrina. Treat all hurricanes as
potentially life-threatening. Regardless of the category, hurricanes move
at 10 to 20 m.p.h. over the ocean. When they move over land, they start
to dissipate, but they're still dangerous for quite a while.

When do hurricanes get names? When they're tropical storms. The
World Meteorological Organization (WMO) has different lists of
names that cycle through the years. (There are different names lists
depending on where the hurricane happens, too.) If a hurricane does
substantial damage, the WMO retires that name from the lists.

Hurricane season in North America is from June 1 through Novem-
ber 30, with the bulk of hurricanes occurring in the fall (after the ocean

has gotten nice and warm). You shouldn't be surprised to find out you're living in a hurricane zone. Hurricanes hit the Gulf states most often, but they can also wander up the Atlantic coast as far north as New York.

Before the Storm

Hurricanes take a few days or even a week to form, so there's a fair amount of warning that something's going to happen.

Boarding Up

The first thing to do to prepare for a hurricane is to protect your windows and doors. A broken window not only lets rain in, which can cause flooding, but wind blowing into the house pushes up on the roof from the inside. If your roof's already a little dodgy, it's possible for the hurricane to peel off the entire roof this way.

The best way to protect your windows is with permanent storm shutters. Protecting the windows is simply a matter of running around the house and closing and latching the storm shutters. Shutters also reduce the cost of most home insurance policies, so there is a direct advantage to having them.

Shutters are most commonly made of aluminum or steel. They cost anywhere from $7 to $35 per square foot, depending on style, material, and strength. Plastic shutters (Lexan or similar), are between $11 and $30 per square foot. You can see through them, but the plastic can fog, scratch, or yellow with age and use.

Hurricane glass is like windshield safety glass, with a plastic layer between two sheets of glass. It replaces the regular glass in your windows, so it's easiest to install as part of new construction. Hurricane glass is expensive, up to $50 per square foot. You can use hurricane glass by itself or you can provide additional protection with shutters.

Another option is to use half-inch or thicker marine plywood that's cut to fit the windows and doors. The individual panels are then screwed down every 18 inches. Plywood shutters aren't as strong as metal shutters and may not meet local building codes. Installing plywood covers is slow, heavy work, taking as much as an hour per window even on a calm day. Their biggest advantage is their cost: between $1 and $5 per square foot.

Life Preservers

Don't bother taping your windows. Tape isn't any protection against the wind throwing something at it. Tape may keep stray shards of glass from flying around when windows break, but it doesn't prevent breakage, which is much more important.

Protecting your doors is fairly easy. Installing solid wood or metal doors may be cheaper than adding storm door coverings. (Solid doors also make your house harder to break into.) The important thing here, if you live in hurricane territory, is to get all this ready to go before there is one. It typically takes about a month to get storm shutters installed, so plan for this in the off-season.

Mobile homes generally aren't as secure as houses, but there are some things you can do before you leave. Make sure that the structure is firmly anchored and that the masonry blocks and foundation are in good shape. Fix siding or exterior panels that may be loose, strap down the roof, and protect the windows and doors.

Making Your Roof Safe

Your roof can be a weak point during a hurricane. The wind pushes up against the roof from under the eaves and peels the roof off. Some roof damage is survivable, but if you lose a lot of roof, your house isn't likely to last. The rain gets into the upper stories and floods the house, weakening the drywall. The wind can then push against the walls and collapse them, and suddenly you don't have a house.

Roofs are most susceptible to damage right where they're attached to the house. Fasten your roof to the frame with straps or additional clips. Clean out your downspouts and gutters so the rain has somewhere to go. Also make sure that the gutters are anchored firmly; a gutter that's peeled off by the wind is likely to take chunks of roof with it and give the wind a toehold to start peeling away still more roof.

Although some roofs are stronger than others, such as terra cotta versus composite shingle, the condition of any given roof can make a profound difference in its ability to survive a hurricane. Inspect the roof at least once a year and after every hurricane.

Securing Loose Items

Anything the wind can move is potentially dangerous. Secure or tie down anything the wind might push around, pick up, or flip over.

- ◆ Bring barbecues inside if possible; if not, anchor them or tie them off to something solid. Disconnect gas lines or LP tanks on gas grills, and bring any LP tanks, charcoal lighter, and bags of charcoal inside.

- ◆ Bring lawn furniture inside or tie it down. If you have a pool, it's standard operating procedure to throw lightweight aluminum or plastic lawn furniture into the deep end. It won't blow away and it's easy to fish out later.

- ◆ Tie down trailers and chock the wheels so they don't roll. Keep motorcycles inside or tie them down under a heavy tarp. (You won't be evacuating on a motorcycle; the danger of getting hit by debris is way too high.)

- ◆ Bolt and strap aboveground fuel tanks firmly to the house. Be sure you know how to turn the fuel valves on and off.

- ◆ Store your generator inside during the storm if you can. You won't be using it until after the hurricane's passed, anyway. If it's too big to move, make sure it's firmly bolted down.

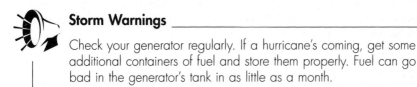

Storm Warnings

Check your generator regularly. If a hurricane's coming, get some additional containers of fuel and store them properly. Fuel can go bad in the generator's tank in as little as a month.

Tidying Up Before the Storm

A major cause of damage during a hurricane is debris that's picked up by the wind and thrown around. You can't clean up everything that's going to get thrown at you, but you can at least eliminate potential missiles in your own yard.

Pick up or secure any loose items: garbage cans, toys, garden tools, wind chimes, lawn ornaments, and dog houses. Nothing will test your shutters as well as having a garbage can or a lawn chair thrown at the house at 100 m.p.h., which is what will happen in a hurricane. Bring all the stray objects into the garage or a storage shed (itself tied down firmly).

Keep your trees and shrubs pruned. Get rid of dead branches that don't need a lot of convincing to fall off and wire down healthy branches that could get ripped loose. Clean up any yard debris that may get flung around. Don't forget to bring flower baskets and small potted plants inside.

Watches and Warnings

When a tropical storm is going to hit an area within 36 hours, a tropical storm *watch* is issued. If the storm is 24 hours away, this is changed to a tropical storm warning. Similarly, when a hurricane is likely to strike your area in the next 36 hours, a hurricane watch is issued. A hurricane *warning* is issued when it's only 24 hours away.

def•i•ni•tion

> A **watch** means that a gale, a tropical storm, or a hurricane is due in 36 hours. A **warning** means that it's due in 24 hours.

A hurricane watch means that unless the hurricane changes its path dramatically or stalls, you're going to have a hurricane on your front porch in the next day or two. Having already made your general preparations, a hurricane watch is the opportunity for you to check everything and do some final specific preparations.

If you live on a coastline, you may see maritime flags that identify the weather conditions. A single triangular red flag indicates a general small-craft advisory for high winds. Two red triangles indicate a gale warning. A single rectangular red flag with a black square in the middle is a tropical storm warning and two of these flags are a hurricane warning. Although Coast Guard stations, lighthouses, and the like always use two flags for a hurricane warning, beaches, boats, and private ships may use just one. No matter, though: the flag means bad weather's coming soon.

Listen to News Bulletins

As soon as a tropical storm or hurricane is anywhere near your neigh-borhood, start checking the weather a few times a day. Storms can pick up strength quickly and their motion is not always predictable. For example, in 2008, Hurricane Fay changed direction and made landfall in Florida four times, something that had never happened before. In addition, if a hurricane is big, evacuation orders can be issued as much as four days before the storm actually makes landfall.

Keep your car gassed up and ready to go. If you need to evacuate, you don't want to be stuck with just a quarter-tank of gas. You may want to have an additional container of gasoline, but gas is highly dangerous if not stored properly, so be careful. It's also a good time to check your flashlights and emergency radios and buy batteries if you need them.

Prep Your Pool

Whether you have an above- or below-ground pool, you'll have a special set of tasks.

- Turn off all lights, pumps, heaters, and other electrical equipment.

- Keep the pool filled. Empty pools can pop out of the ground due to an excess of groundwater from rains or flooding. (This is true for a while after the hurricane's over, too, so don't drain the pool right after the storm, either.)

- Close the valves and remove the pumps, or wrap the pump motor with waterproof plastic and tie it in place. Do the same thing for time clocks, heaters, and light transformers. If it looks as though there's going to be flooding, you may want to disconnect the equipment and store it inside.

- Shock the pool with extra chlorine to prevent contamination.

- Keep the pool cover off. The wind can pick up the pool cover and rip it or send branches through it. It's easier and cheaper to fish debris out of the pool after it's all over.

Prep Your Boat

Boats and hurricanes don't do well together. If your boat's on a trailer, keep it in a covered storage area. If covered storage isn't available, find a protected area, let the air out of the trailer's tires, and lash the boat and the trailer down. If you can, tie off to heavy fixed objects in four different directions to anchor the boat in place. Small boats can be filled with water after you lash them down just to give them more weight.

If your boat's in the water:

- Stow all loose gear.

- Add old tires or other crash protectors to the sides of the boat.

- Guard against chafing with chafing sleeves and double up all your lines.

- Make sure that your hardware is lashed down, bolted down, or otherwise secured.

- Charge the batteries on the automatic bilge pumps.

- Seal all the openings with duct tape.

For added security, remove the electronics and outboard motor from the boat and store them.

The Final 24 Hours

Whereas a hurricane watch is a suggestion that you should start getting ready, a hurricane warning is the equivalent of saying "… and we're not kidding!" A hurricane warning is issued when hurricane conditions are expected in 24 hours or less. By this time, the weather's pretty icky and it's clear that you're in for trouble soon.

Most of your preparation should be done by the time a hurricane's about to hit, but there are a few tasks to save for the last minute:

- Check that the shutters for your windows and doors are latched in place.

- Fill your water jugs or bathtub with fresh water. Safe drinking water isn't always available during a hurricane because of broken pipes and contamination.

◆ Do a final check of the house and yard.

◆ Turn your refrigerator and freezer to their coldest settings, and open them only when necessary and for the shortest possible time. Turn off any other appliances you don't need to reduce the load on the local power grid.

◆ Check next door and across the street to make sure that they're all ready, too. Not only is this being a good neighbor, you can make sure that they're tied down and cleaned up so you don't get hit by their garbage cans.

◆ Keep monitoring the news.

Evacuation Issues

If your house is in good shape and outside the main storm zone, and you haven't been told to evacuate, stay put. Evacuating when you don't need to adds traffic and may put you at greater risk of being stuck on the road instead of in a nice, dry house. However, any house or building on the coast or a floodplain, or near a river or inland waterway, is in danger from storm surges.

If you live in an apartment building or mobile home, it's safest to evacuate. Even if the apartment building has storm shutters, the storm winds get stronger the higher up you go. There's also likely to be more debris flying 40 feet off the ground. If you can't evacuate, take shelter as close to the ground floor as you can.

Regardless of where you live, be ready to evacuate the moment local authorities say to evacuate. Leave early and avoid the rush. The weather may actually still be good when the evacuation order is issued, but leave immediately anyway. You may need all that time and more to get to safety. You're not the only person leaving the area, after all.

◆ Dress warmly in layers. You can carry more clothes this way.

◆ Pack what you need. Load your Go-Paks, emergency supplies, blankets, and sleeping bags in the car.

◆ Lock the house and hit the road.

Follow the evacuation directions given on the news. Don't take short-cuts; they may not be safe. Keep the car radio on in case there's a change in road conditions or evacuation.

Shelter in Place

The storm may be coming and the best thing would be not to be there, but you might not be able to evacuate for a number of reasons, such as ...

♦ You have pets or livestock that you can't evacuate.

♦ You or a family member have medical problems that make it difficult or impossible to travel.

♦ You don't have adequate or reliable transportation.

♦ There are too many people on the roads already or the storm has cut off the escape routes and evacuation is impossible.

For long-term storm planning, you can build a heavily reinforced safe room. (You can also buy pre-fabricated safe rooms that are basically large steel boxes.) You won't have a lot of room in a safe room, but they're all about safety, not comfort. A storm cellar may be safe, but cellars can flood, particularly if there are storm surges. Don't plan on using a storm cellar for refuge unless you know it's safe from flooding. You don't want to get flooded out of your safe place at the height of the storm.

During the Storm

You're as ready as you can be to ride out the storm. As the hurricane progresses, check the inside of the house periodically. Look for leaks or wet spots, which are signs that rain is pouring in from a weak point in the wall above or the outside, meaning that the roof may have developed a leak or the siding has come off and water is being forced in. Also pop your head into the attic to make sure the roof is still attached firmly. (Avoid going into the attic; if the roof pops off at that moment, you'll get blown off the top of the house.)

◆ If there's no power, use flashlights or glowsticks for light. Avoid candles or open flames: the risk of fire from a stray draft is too great.

◆ Keep monitoring your radio or TV for weather advisories and emergency information. The hurricane's already hit, but there can still be important announcements, such as a late call to evacuate, a change in the hurricane's direction, or collateral storm damage or hazards.

◆ Close all the doors in the house and brace the exterior doors. Stay away from windows. No matter how well your windows are shuttered or reinforced, there's always a chance the wind will throw something at you (such as a shed) that's big enough to break the glass and hurt you.

◆ Stay indoors on the downwind side of the house. Stay in a small interior room, a closet, or a hallway or bathroom on the lowest level of the house, or lie on the floor under a table. Use mattresses for protective padding.

 Storm Warnings

If the hurricane's eye passes over you, you'll have a short period of calm weather, after which the winds will come at you from the other direction. Stay inside; when the wind starts picking up again, it'll go from zero to hurricane force very quickly.

Related Hazards

Hurricanes bring high winds and rain, bad enough by themselves, but they also come with the potential for significant collateral damage.

Flooding and Storm Surges

Hurricanes can easily produce more than a foot of rain, which can saturate the ground, raise the level of local waterways, and cause flooding. Check with your city or county emergency services agency or your insurance agent to find out if you're in significant danger from floods.

In addition to the rain coming down, a storm surge is an excessive amount of water that's pushed inland by wind. Storm surges usually cause the largest number of deaths in a hurricane.

Prep Facts
Don't drive into water of unknown depth across a road. What looks like a small stream may actually be a deep ditch with a rapid wall of water that can sweep your car away and drown you.

If you're at risk for floods or storm surges, you may need to be prepared to protect your house from high water. Keep sandbags, plastic sheeting and garbage bags, plywood and lumber, shovels, and work clothes for both hazards. You may also want to move furniture and valuables to a higher floor to keep them from being damaged.

Tornadoes

Being hit by a tornado during a hurricane has an "insult to injury" feel to it, but a hurricane can suddenly spawn tornadoes on its outer edges. Despite being much smaller, a tornado can be much more destructive on a local basis than a hurricane. See Chapter 9 for detailed information on tornadoes and how to survive them.

In addition to damage to your roof and windows, the wind can fling branches, knock over trees and utility poles, and rip out power lines. During a storm, your best bet is to stay inside where you're protected.

Lightning

Hurricanes can spawn terrific lightning storms as they travel. Staying inside houses or cars with metal roofs is safe. Standing under or near trees, metal poles or fences, tents, picnic shelters, or water is not. If you're caught in the open, try to make yourself as small a target as possible. Don't climb on the roof when there's lightning; you're just auditioning for a role as an amateur lightning rod.

Not So Calm After the Storm

Usually after a few uncomfortable days in a shelter or holed up in your house, the hurricane has dissipated and the flood waters have receded.

Now the sky is blue, the grass is green—and lots of hidden dangers may be lurking. If this is the first hurricane you've weathered in this location, you may discover that you need to tie things down differently or that a tree needs to be pruned heavily before the next storm.

Some of the most common hazards after a hurricane are broken electrical, water, and natural gas lines; damaged or weakened houses and buildings; stagnant, toxic water; and contaminated water supplies. Work with the local authorities (who are probably spread too thin) as best you can, and be patient.

Storm Damage to Roofs and Houses

After the storm clears, do a careful inspection of your house for damage. If you're not adequately shuttered, you'll probably have at least one broken window or door.

Lifting of the roof can be very serious, because it weakens your roof for the next hurricane and gives water a way into the walls. Don't be surprised if you've also lost shingles or roofing. Roof turbines, antennas, and satellite dishes are a frequent casualty.

Because they're so large, garage doors can buckle or jam. In addition, they may take damage from debris thrown around by the wind. Debris may also have left dents or other damage all over the house. In fact, an initial impact during the storm can loosen the siding enough so that a piece comes off. After that, the wind can peel the siding off the house piece by piece.

Basements, crawlspaces, and anything below ground level is prone to flooding. Although many sump pumps work underwater, the power may be cut off. Sump pumps also aren't effective if the ground is saturated or if the water is coming in faster than the pump can get rid of it. If your sump pump isn't working, consider using a marine bilge pump to drain the worst of the water.

Finally, be on the lookout for fire damage. All the water and moisture can saturate wiring and cause shorts, which in turn can start fires when the power's back on and things are beginning to dry out.

Drinking Water Safety

Unsafe water is a major hazard after the skies clear. If there was flooding or a storm surge, assume that the flood waters are contaminated with sewage and toxins:

◆ If you have well service, pump the well out and then test the water.

◆ Discard any fresh food that came in contact with flood waters.

◆ Use soap and hot water to wash cooking utensils and canned goods that have been contaminated with flood waters.

The Least You Need to Know

◆ Hurricanes and their aftermath can last for up to a couple weeks. Stock emergency supplies accordingly.

◆ Whenever a hurricane is approaching, keep listening to radio and TV bulletins.

◆ Clean up and/or tie down anything that could get picked up by winds. Secure your windows, your roof, your car, and your boat.

◆ Hurricanes can stall, change direction, and pick up speed in a very short time. Be flexible; it may be safe to shelter in place one minute and necessary to evacuate the next.

◆ Most homeowner's insurance does not cover wind damage and no homeowner's insurance covers flood damage (groundwater). Review your coverage yearly with a qualified insurance agent to protect your home and your family.

Chapter 9

Tornadoes

In This Chapter

- How to know if there's a tornado in your area
- What to do and where to shelter when a tornado strikes
- How to safely shelter in place
- How to stay safe after a tornado

Of all the natural disasters, tornadoes are the most capable of creating sudden, widespread devastation. Floods and hurricanes cover larger areas, earthquakes are even more abrupt, and blizzards last longer, but tornadoes can hit hard and fast, with little or no warning. When it comes to tornadoes, disaster preparation during the "good times" is key to ensuring the safety of you and your loved ones when a tornado strikes.

In this chapter, we discuss what tornadoes are and what causes them, what they're capable of, where and when they occur, how to shelter from them, how to read the warnings, and how to take cover. We'll also tell you about tornado-related hazards and how to deal with the aftermath.

What Is a Tornado?

A tornado is a violent, rotating column of air underneath a cumulus or cumulonimbus cloud (a thunderhead) that also touches the ground. Beyond that, the distinctions get a little fuzzy: tornadoes don't all look like something out of the *Wizard of Oz* (most of them don't, in fact). A tornado usually has a visible funnel, but not always. There is frequently a cloud of debris around a tornado's base, but not always. Tornadoes can be tall, thin, and very sharply defined or they can look like large storm fronts with bases up to a mile wide. They can even have multiple funnels. There's just no telling.

The most frequent cause of tornadoes is along weather fronts, often where masses of warm, moist air collide with colder, dryer air. If the lower layer of air moves upward, a spiral movement can start that convects more air up and increases the strength of the spiral winds. This will almost certainly generate thunderstorm activity and it may also cause a tornado to form. However, tornadoes can form in other ways. For example, tornadoes can form out of supercells, which are large, rotating thunderstorms. Imagine the typical thunderhead, but because it's rotating, the cloud formation is rounded.

Whatever the cause, when a tornado begins to form, there's usually a funnel cloud. Sometimes the funnel is visible coming down from the parent clouds, but the funnel cloud isn't always visible at first. As a result, the first sign of a tornado is often the whirling cloud of debris that appears suddenly when the column of rotating winds hits the ground. Dust, dirt, and moisture get sucked up the column and then the tornado is visible.

Prep Facts
Tornadoes can be many different colors, largely depending on what's being pulled up into the funnel. Usually gray or brown from dirt, tornadoes can also be red in red plains areas, green when they're hitting lots of vegetation, soft white over water, and even brilliant white when they're traveling over snowfields.

Tornadoes have a strength classification system known as the Fujita-Pearson scale, or simply the F-scale (similar to the Category 1 through

Category 5 grading system for hurricanes). Tornadoes can be F0 through F5 depending on their windspeed and the type and quantity of damage they can do to a well-built house or building. However, in 2007, the Fujita-Pearson scale was superceded by the Enhanced Fujita scale, abbreviated EF0 through EF5. The EF scale is similar to the F scale, but it evaluates tornadoes more on the basis of damage than on windspeed and adds more structures and vegetation as ways to measure the damage:

- **EF0:** Windspeeds of up to 85 m.p.h. Light damage to chimneys, roofs, gutters, and siding. Trees lose branches and shallow trees may be uprooted.

- **EF1:** Windspeeds of 86 to 110 m.p.h. Moderate damage to structures, such as garages being flattened, roofs being stripped, windows and doors broken or peeled off, mobile homes knocked over, moving cars pushed off roads.

- **EF2:** Windspeeds of 111 to 135 m.p.h. Considerable damage to buildings, mobile homes destroyed, large trees uprooted or broken, cars lifted off ground, boxcars overturned, light objects becoming missiles.

- **EF3:** Windspeeds of 136 to 165 m.p.h. Severe damage to houses, buildings, and shopping malls, mobile homes and structures with weak foundations blown away, trees in forests uprooted, broken, or debarked, trains overturned, heavy cars lifted off ground and thrown. Small objects becoming missiles. Serious injury or death is very likely at this tornado strength and higher.

- **EF4:** Windspeeds of 166 to 200 m.p.h. Devastating damage to houses and buildings, houses leveled, cars thrown, large objects becoming missiles.

- **EF5:** Windspeeds of more than 200 m.p.h. Incredible damage— general destruction to houses and buildings. Strong houses lifted off foundations and blown away, automobile-sized objects becoming missiles with distances of more than 100 meters, steel-reinforced concrete structures badly damaged, high-rise buildings are significantly structurally deformed. Incredible phenomena occur.

Of the more than 1,000 tornadoes in the United States annually, roughly 85 percent are EF0 or EF1; another 10 percent are EF2. Only 5 percent of tornadoes are EF3 or higher, but they account for 75 percent of tornado deaths. EF5 tornadoes are very rare: fewer than one EF5 tornado a year has occurred since 1950, but they have also been responsible for more than 1,000 deaths.

Most tornadoes don't last very long. EF0 and EF1 tornadoes usually last fewer than 10 minutes. EF2 and EF3 tornadoes can last for 20 minutes or more. EF4 and EF5 tornadoes can last more than an hour, adding to their destructive potential. Sizes and lengths of the destruction path vary tremendously, too, from only a few feet wide and a few miles to a couple miles wide at the base traveling across counties and even states. (One record tornado in 1925 was on the ground continuously for 219 miles and traveled through three states.)

Touring Tornado Alley

Although tornadoes can happen anywhere, they occur most frequently in the United States and southern Canada. This is largely a function of the terrain, as the large plains areas in the middle of North America have weather conducive to tornado formation. Tornadoes can also happen as part of a hurricane or tropical storm front, forming ahead of and to the right of the storm center.

Tornado Alley is the nickname commonly given to the states most likely to get hit with tornadoes: Texas, Oklahoma, Kansas, Nebraska, and Iowa. Neighboring states, such as South Dakota, Minnesota, and Colorado, have frequent tornadoes, too, but not usually with the intensity of Tornado Alley.

Tornadoes in the United States can happen at any time of year, although they are more common in the winter and early spring in the Southeast. Spring and summer see the most tornadoes happening in the central United States and, as the weather gets warmer, southern Canada.

Tornadoes are most likely to happen in the late afternoon and evening, but they can happen at any time. In general, tornadoes move from the southwest to the northeast, but they can move in any direction. The speed of a typical tornado is about 30 m.p.h., but they can move at just a few m.p.h. or as fast as 70 m.p.h.

Storm Shelters and Safe Rooms

Although most other disasters allow for the idea of evacuating to another locale, tornadoes tend to be a "shelter in place" disaster. You rarely have enough warning to get away and it's not always easy to tell where you should escape to. Although tornadoes usually move in a fairly predictable way, they can change direction without warning. Worse, it's possible for a second tornado to touch down without warning, meaning you could suddenly find yourself driving straight into a funnel cloud. As a result, the best way to stay safe when a tornado passes by is to shelter in place.

It's prudent to have shelter locations in both your home and business, either a storm shelter or a safe room. Storm shelters are underground structures designed to withstand the force of a tornado and to provide shelter from any debris the tornado may throw at it. Traditional storm shelters were basements or root cellars with a heavy door that could be barred from the inside. If a tornado passed right overhead, it might damage the house or building above, but it's unlikely that it would harm the people inside. Storm shelters today tend to be small rooms built into the foundation with a small access hatch. Even in areas with high water tables, you can get fiberglass, steel, or lined concrete shelters that seat up to a dozen people and are built completely flush to the ground. If you want to build something yourself, most communities in tornado zones have information on approved aboveground shelter styles.

You may not be able to build a belowground storm shelter because of zoning requirements or utility pipes, but you can always build a safe room. A safe room is a durable box, usually steel or steel reinforced concrete, which is anchored very securely to the foundation slab. You can purchase safe rooms as prefabricated units. An advantage of a safe room is that they are wheelchair accessible, where storm shelters frequently require access through a short, rather steep staircase. In addition, safe rooms tend to be taller than storm shelters, which are frequently designed for you to remain seated. This may not be a problem, however, as you usually aren't in a storm shelter or safe room for more than 10 or 15 minutes at a time.

Life Preservers

Should you go for a below-ground or aboveground shelter? It's always safer to be below ground during a tornado, where you're not likely to be hit by debris or caught in a tornado's winds, but digging a hole is more expensive and it may be impractical if you have bedrock or a high water table. Aboveground shelters are easier for seniors or disabled people to get into, but they take up space and are somewhat less secure. Both types of shelters are more difficult to build into a pre-existing house, but having them inside your home means you don't need to lock them to keep their contents safe. The 30 seconds it takes to find and use a key if a tornado is coming may be all the time you have.

Warnings

The National Weather Service (NWS) has a fairly sophisticated network of Doppler radar stations centered in Norman, Oklahoma, called NEXRAD (short for "NEXt-generation RADar"). These stations cover 98 percent of the continental United States. They scan for severe weather conditions and provide alerts of potentially dangerous storm activity.

Although NEXRAD can detect the weather conditions that are likely to produce tornadoes, radar can't always spot a tornado as it's forming. For this reason, the actual identification of tornadoes is usually done through SKYWARN: an NWS network of almost 280,000 volunteer storm spotters that observe local severe weather conditions and report on them to the NWS offices. Storm spotters include police, sheriff, and fire departments, emergency personnel, amateur radio operators, and citizen volunteers.

When conditions are likely for tornadoes to form, a tornado watch is issued. The storm spotters in a tornado watch area are alerted to watch for tornadoes and other unusual severe weather conditions. If you're at home during a tornado watch, keep the radio or TV on to listen for weather bulletins. Have everyone in the family keep their Go-Paks handy—when a tornado touches down, you may need to literally grab and go. Move your cars into the garage or carport and leave them unlocked for fast access. Have house keys and car keys with you at all times. Also, if time permits, move anything the storm could throw at the house inside, such as lawnmowers, garden tools, and patio or pool furniture.

If conditions worsen, a tornado warning is issued. This means that a tornado is imminent (because of a visible funnel cloud or a rotating thunderstorm that is likely to produce a tornado) or a tornado has actually been sighted.

 Storm Warnings _____

Tornadoes come with a variety of warning signs. One of the most commonly reported is the sky turning a sickly green or yellow color. This doesn't happen with all tornadoes nor does it always mean a tornado is about to happen, but if you see the sky looking green or yellow, check out the weather reports immediately. Other danger signs for tornado conditions include unusually large hail or a solid line of clouds that looks like a wall.

In extreme conditions where an exceptionally large tornado is likely to hit a populated area, the NWS may declare a tornado emergency. This is a weather bulletin that happens in addition to a tornado warning and anyone in the affected area should take immediate shelter.

Sirens

Many communities no longer have the bright yellow warning sirens on telephone poles that were installed as part of the civil defense programs of the 1940s and 1950s. However, towns and cities in tornado areas may still have tornado warning sirens or alarms. Check with the local fire or police department to find out what general alert systems are likely. Many communities have begun to install sirens again based on the need to reach the general populace who may not be near a phone, radio, or TV.

One of the best ways to monitor weather conditions is to buy a weather radio. This is a small unit that's always turned to the emergency frequencies used by the National Oceanographic and Atmospheric Administration (NOAA). When there's a weather alert, there is an alarm followed by a broadcast. Most weather radios for sale also allow you to replay the last weather alert message that came in, in case you missed it for whatever reason.

Tornadoes may also trigger the reverse 911 system: an automated warning is phoned out to every phone number in the coverage area. This warning may be of limited effectiveness because of damage to the phone lines caused by a tornado in the area as well as the limited warning time with many tornadoes. There also isn't any way to replay a message from the reverse 911 system.

Radio and television provide very detailed and immediate information through local weather alerts and the Emergency Broadcasting System about tornadoes and weather conditions. The only problem with this is that these alerts depend on people having a radio or TV on when the bulletins come in.

Life Preservers

Of all the tornado warning systems available, weather radio may be the best. You don't have to have the radio or TV on, they'll alert you even when you're asleep, there's a message recall option, and they're immediate. As anyone who lives in a tornado area can tell you, everything stops when the weather radio goes off.

Seconds to Shelter

Of all natural disasters, tornadoes are second only to earthquakes for short warning time. Tornado watches and storm spotters can provide some advance warning, but tornadoes can and do form at night when it's not possible to see the conditions that lead up to them. Even during the daytime, the funnel may not be visible. All you may see is a cloud of debris. If there are buildings, trees, or terrain between you and the tornado, this may not be visible until the tornado is almost on top of you.

Another sign of a tornado can be deceptive: the winds may die down and the air becomes very still. This can also be the sign that a tornado has dissipated, so it's not always clear that the tornado is still coming. But because tornadoes usually happen at the trailing edge of a thunderstorm, it's not surprising to see a tornado with clear skies behind it.

No Time to Evacuate

In some cases, the first sign that there's a tornado is when it's already touched down nearby. You may hear the tornado nearby—often described as sounding like a waterfall, jet engine, or locomotive—or you may hear the sound of debris hitting your house. These mean you have only seconds to find shelter. Be ready to head for shelter immediately.

If you're in your home or other building, head for the storm shelter or safe room. Failing that, go to the basement or lowest floor. If neither of these is possible, move to an interior room or a hallway on the lowest floor you can reach. Even a closet will do. The trick is to put as many walls and doors between you and the tornado as you can. Get under a sturdy piece of furniture, such as a kitchen table or a workbench. Keep your head and neck covered. If you have time, you can use a mattress to provide additional shielding from flying debris. Stay away from windows, which provide no protection from anything thrown by the tornado and will shower you with shards of glass if they break.

Mobile homes are very difficult to anchor securely and they're not safe. Even an EF1 tornado is capable of knocking them over and an EF2 tornado can destroy them entirely. If you're in a mobile home, use what time you have during a tornado watch to get to secure shelter.

If you're driving and you see a tornado in the immediate vicinity, don't try to outrun it. An EF1 tornado can push a moving car off the road and an EF2 tornado can pick a car completely off the ground. Instead, get out of the car and seek shelter. If you're not able to get to shelter, lie face down in a low area and cover the back of your head and your neck with your hands. *Never* take shelter under an overpass! The narrow channel under the overpass can bottleneck the winds, causing them to speed up tremendously. People have been plucked from under overpasses by the extreme winds and killed in cases where there were no other fatalities for that tornado.

Shopping malls and large stores in tornado-prone areas usually have a tornado shelter. (If there aren't signs for a tornado shelter, look for the yellow-and-black Civil Defense shelter signs.) If nothing else, head for the lowest floor and the most sheltered room or area. Stay away from large open rooms and windows.

If you're out in the open when a tornado touches down, move at right angles from the tornado. Although tornadoes can change directions, they will generally continue in the same direction they're moving. If you can't reach shelter (such as a building with a basement), lie flat in a ditch or a depression. This will keep you from being struck by debris flying at or above ground level.

Avoid taking shelter near outside walls, windows, or elevators, in areas that may flood, in cars or trucks, or in buildings that have large roofs (which are susceptible to being peeled off by strong winds).

Related Hazards

Tornado winds are very dangerous by themselves, but there are additional dangers you need to be prepared for.

Flying Debris

Most of the deaths and injuries from tornadoes are actually caused by flying debris. Tornado winds can fling objects of almost any size at speeds in excess of 200 m.p.h. A 6-foot 2×4 moving at 80 m.p.h. can go through a solid wood exterior door. The same 2×4 moving at 150 m.p.h. can go through an 8-inch concrete block wall. Small objects, such as nails, can cause massive penetrating injuries at tornado speeds. (By comparison, home nail guns only shoot nails at 70 m.p.h.) And even if there's nothing else for the tornado to fling, tornadoes are capable of producing their own projectiles in the form of hail.

Falling Flotsam

Tornadoes pick things up in their funnels and push them high into the air. How high depends on the weight of the object and the strength of the tornado. (Some tornadoes are capable of pushing smaller objects up into the lower stratosphere at 35,000 to 40,000 feet, where jets typically fly.) When the tornado dissipates or when the object is ejected from the tornado's funnel, it falls to Earth. This can be anything from large hail to debris from shattered buildings to rocks and branches to fish. If the tornado has passed over water, it picks up large volumes of water and anything in it. The water dissipates in the funnel, but the fish, frogs, or whatever is deposited over land when the tornado dissipates.

Even when materials haven't been lofted to the stratosphere, there's still the problem of structures losing windows, facades, or falling apart completely because they've been rendered unstable by a tornado.

Lightning

Tornadoes are usually created by storm clouds, storm fronts, or similar kinds of severe weather. Accompanying this kind of weather are lightning, hail, and rain.

Lightning kills about 75 people every year, four fifths of them men. If you can see lightning or hear thunder, you're potentially in the strike zone for the next lightning strike. Seek shelter in enclosed buildings. If you're outside, crouch or lie down. Stay at least 7 feet away from isolated tall objects such as trees or poles and get rid of large metal objects.

Hail from tornadoes and the accompanying storms has been known to get as big as grapefruits, and hail causes several hundred million dollars in damage annually to property and crops. Although hail rarely causes injury or death because big hail is rare and people can take cover when it's hailing, it can happen. For example, there are several cases of small planes being brought down because of large hail.

The rain associated with tornadoes can be sudden, hard, and plentiful. Tornadoes can dump extremely large amounts of water in places that may not get a lot of water, which can cause flash flooding. Again, this may be on top of the large quantities of rain that surrounding storm clouds might have dumped.

Following the Funnel

Except in the rare cases of EF4 and EF5 tornadoes, most tornadoes will be over in 15 minutes or less. If you've taken shelter, you're going to be safe. Now you've got to deal with the aftermath.

If the power lines are aboveground or attached to a structure such as a house or building, there's a risk they'll be exposed and possibly torn free by a tornado. In addition, because there's not necessarily going to be water on the ground, downed power lines are not necessarily going to ground out and will remain live. Always treat all downed cables of any kind as potentially live and lethal.

Most gas and water lines are going to be underground. However, they can be broken where they approach or enter a mobile home, house, or building. If a structure is ripped loose from its foundation or flattened by a tornado, gas and water pipelines are going to be severed, releasing combustible gas and water. This can result in fires, explosions, or water damage to what's left of the building.

Possibly the most common type of structural damage caused by tornadoes is the partial or complete loss of roofs and top levels of buildings while leaving the rest of the structure intact. This can be caused by winds pushing at exposed corners of roof shingles and peeling off sections of roof or, in more extreme cases, by flying debris causing initial damage that the tornado winds can then exploit further.

The Least You Need to Know

- Although some states are more prone to tornadoes than others, tornadoes can happen anywhere, any season, any time of day.

- Sheltering from tornadoes requires some planning, but it's worth it: you can't evacuate from a tornado like you can with a flood or hurricane.

- Tornado alerts depend on sophisticated radar as well as a huge network of volunteer storm spotters.

- A tornado's winds are dangerous but not nearly as damaging as the things they can throw.

- After the tornado is over, watch out for downed power lines and broken gas and water pipes.

Chapter 10

Wildfires

In This Chapter

- ◆ Determining where and when you're likeliest to encounter wildfires
- ◆ Fireproofing your house as much as possible
- ◆ Avoiding starting a fire yourself
- ◆ Dealing with a fire in your area and knowing when to evacuate
- ◆ Making a faster recovery by understanding fire

The image of a quiet campfire or a romantic fireplace are one aspect of fire in our lives. Fire cooks our food, even adding flavor. Fire heats our home and even adds romance to an otherwise mundane evening. But fire has another side, unbridled and ferocious. Wildfire not only sounds ominous, it is one of nature's most frightening disasters.

In this chapter, we describe wildfire and the many factors that make certain regions of the country more prone to wildfire than others. We then guide you through the process of mitigating your risk and creating a fire-preventive landscape around your home. After you are prepared, we help you determine when to evacuate and what to take with you given the much higher than usual likelihood of losing treasured mementoes in a fire. Finally, we prepare you for the aftermath of a wildfire both physically and emotionally.

What's a Wildfire?

A wildfire is any unwanted fire that burns scrub brush, grass, or forest. More importantly, a wildfire is any fire, either naturally occurring or man-made, that is at some point uncontained in a natural habitat.

One of the reasons for many wildfires is, ironically, fire suppression. Small forest or brush fires that would normally have been started by lightning and would burn themselves out fairly quickly with minimal damage are now stamped out quickly. As a result, larger accumulations of flammable material build up, so when a fire finally does break out, it has a lot of fuel, is much hotter, and causes much more damage.

Hand in glove with this is the problem of more houses being built in or next to potential fire areas. Where small brush fires would have burned themselves out in the past, they now can endanger whole housing developments. To complicate all of this, drier, hotter weather as a result of climate change and overuse of local water tables have both contributed to longer, more intense dry seasons.

Regardless of the fuel, wildfires are almost always associated with winds that spread the fire by blowing sparks and embers to new locations. Sparks generally fly only half a mile to a mile, but there are many cases of fires being ignited by embers from 6 miles away.

Brush Fires

Brush fires generally occur in late summer and fall in arid and semi-arid climates. Creosote chaparral, sagebrush, and other desert plants are the most frequent fuel source in wildfires in Southern California. Prairie grasses, shrubs, and dry swamp grasses are fuels for Midwestern and Southern fires. These plants can dry out enough to burn even in normal years. Drought conditions can exacerbate this tremendously, causing large quantities of brush or grasses to dry out and making them susceptible to uncontrolled burning.

Another problem is the introduction of non-native plants to an area. The local plants may have struck a balance with fire conditions, but any new plant that grows like, well, a weed can create a new fire hazard by adding prolific amounts of new fuel.

Forest Fires

Forest fires are also associated with hot, dry conditions. There are two types of forest fires. Surface fires burn relatively quickly and consume the stray brush, dry limbs, and leaves. Because surface fires move fairly quickly, they cause relatively little damage to the forest and are actually an essential part of the forest lifecycle.

However, if there is too much fuel because conditions are exceptionally hot or dry or the trees are stressed from insect or fungal infections, a much more dangerous crown fire can occur. Crown fires burn the forest canopy, seriously damaging or killing the trees. Crown fires are also much more likely to spread to other areas because of the quantity of burning material and the updrafts created by the fire itself.

Dry in the High Chaparral

People like to build houses in places that are either near oceans, lakes, and rivers and therefore prone to flooding, or near forests, on mountains, or in deserts, where they're prone to burning. Some of the most expensive homes in wildfire zones are built in the high chaparral, which are some of the most flammable places.

One of the most important issues with wildfires is the fuel source. Obviously, dry material burns better than wetlands, but many areas in the western United States contain naturally growing chaparral, which burns as part of its natural life cycle. Chaparral is especially flammable and burns at a high temperature due to resins in the plant and leaves. Similarly, many species of pine release their seeds from cones only after exposure to wildfire. As with the chaparral, these pine species are especially flammable.

The fuel is not the only factor, however. Tree-top burns (king fires) and swamp/lakeshore fires (muck fires) pose special dangers because of the speed with which they spread and the difficulty in containing and extinguishing them. Unlike ground fires, king fires jump from tree top to tree top. They can spread through an enormous area very quickly because of the strong updrafts and the flammability of the tree tops as they dry out and heat up from the hot winds created by the fire. Similarly, muck fires can burn silently underground and re-erupt a long way from the original outbreak hours or even days later. Detecting and extinguishing underground fires is very difficult.

Before the Blaze

You may not be able to prevent a wildfire in your area, but you can do a lot of things to help fireproof your house and your neighborhood.

Create a safety zone that contains no supplies of fuel that are easily ignited. You want to reduce the risk of flames and heat. How big the safety zone needs to be depends on where you are. If your house is on level ground, a safety zone of 30 feet may be adequate. A house in a forest requires a 100-foot safety zone, because of the larger amounts of heat and the possibility of burning trees falling on or near your house.

Regular cleaning and maintenance will also reduce your risk of fire:

- Keep the roof and gutters cleaned. Don't let leaves or pine needles build up.

- Clean your chimneys annually and use approved spark arresters.

- Use a ¼-inch mesh grill screen to prevent stray embers.

- Eliminate stacks of newspapers and flammable rubbish. Use appropriate containers for storing gas and flammable liquids.

- Dispose of ashes from barbecues, fireplaces, and wood stoves by leaving them in a metal container with water for a day or two to make sure there are no stray embers.

- Keep stacks of firewood covered and at least 100 feet away from your house. If you're on a slope, the firewood should be uphill: if it catches fire, the heat and embers will move away from your house.

Cutting Back Brush

The first thing to do to create a safety zone is to cut back brush and grasses that can burn easily. Rake leaves, twigs, and dead tree limbs and dispose of them safely. Don't let leaves and flammable rubbish accumulate under or near your house or nearby structures, either. Keep the grass mowed.

If there are trees near your house, prune any dead branches near your house as well as any branches within 15 feet of the ground. Thin branches to allow a 15-foot gap between the crowns of trees so fire can't spread easily from tree to tree. Don't cause a fire, either: remove branches and shrubs that are within 20 feet of a chimney outlet, and alert your power company to branches that are in or dangerously close to power lines.

The tools you use for maintenance are good for fighting fires, too: rakes, shovels, buckets, axes, and saws. You also should have a ladder tall enough to reach the roof.

Preventive Landscaping

It won't do you any good to create a fire safety zone if you then landscape with plants that burn easily. When you're planning your landscaping, use fireproof or fire-resistant plants.

- Separate trees and shrubs with walkways, rocks, or other areas without fuel by 10 to 15 feet.

- Eliminate weeds such as thistles and prairie grasses that can burn easily.

- Plant fire-resistant trees such as big-leaf maples, aspen, and oaks, and fire-resistant shrubs such as honeysuckle, grapes, and viburnum. (In general, hardwoods and fruit trees are safer than evergreens and eucalyptus.) Mature trees, particularly those with heavy bark, are much less flammable than saplings.

- Replace wooden structures with stone, concrete, or other noncombustible materials. Treat exposed wood with fire-retardant chemicals. Move flammable objects, such as wooden benches, away from the house.

◆ Remove plants and other combustibles within 10 feet of propane tanks and gas barbecues.

Check with a local nursery about what kinds of plants are noncombustible.

What's Wet Doesn't Burn

Fires require three things to burn: oxygen, heat, and fuel. If you remove one of these three things, the fire will go out. Water is effective for putting out fires because it prevents oxygen from reaching the fuel and it cools the fire below the temperature necessary to continue combustion. Wet fuel won't burn because the water continues to block the oxygen from the fuel and it also wicks away heat, keeping the temperature down.

As part of your fire-prevention efforts, identify sources of water that you may be able to use: swimming pools, wells and cisterns, or even a small pond or lake. In urban settings, locate the fire hydrants nearest your house. Get a pump that allows you to draw water directly into a hose. Many people have successfully prevented their houses from burning during wildfires by pumping water from their swimming pools into hoses and keeping the house and surrounding vegetation damp.

Have a garden hose long enough that you can reach any part of your house. (You may need several hoses to accomplish this.) Also make sure that you have enough water outlets and that they're conveniently located on several sides of the house. It's a good idea to have water outlets 50 feet from your home, as well: if the house is on fire, you may not be able to get to a hose bib on the side of the house, but you can use a nearby outlet to get as close to the house as the heat allows.

Fireproof Materials on New Construction

If you're building a house or other structure, use fireproof materials. Many municipalities have requirements about treated or fire-resistant materials and fire-prevention devices (such as spark arrestors on chimneys). Look into ways to provide additional fireproofing, such as using concrete or noncombustible plastic materials. In particular, look into noncombustible siding and roofing materials.

Smoke Signals

Most wildfires are caused by people, either by accident or intentionally. In addition to the fire-prevention techniques mentioned earlier, some other things you can do to avoid starting fires include:

◆ Watch for stray sparks from a chimney or barbecue.

◆ Don't park your vehicle on dry grass. The grass can ignite from touching the exhaust manifold. The heat may also ignite the car, providing a large amount of flame and fuel to the surrounding area.

◆ Obey all fire regulations and burn bans.

◆ Use spark arresters on any off-road vehicle.

◆ Discard cigarette butts and pipe ashes in metal containers or in water. Never just throw them outside.

◆ Keep any outdoor fires small and in a tightly confined area. Always have a shovel and water on hand in case the fire ignites something nearby. Never take burning material out of a fire.

◆ Make sure that fire vehicles have an unobstructed path to your house. Prominently mark your address and your driveway entrances to make your house easier to find for fire and emergency crews.

◆ Don't let children play with fire.

◆ Never leave a fire unattended!

If you are in an area that is currently at risk because of dry conditions, check news bulletins regularly. Wildfires are going to be reported quickly because they're news and people need to know about them.

Fires can move very quickly. If a wildfire is burning in the general area, keep track of bulletins on the fire's direction and proximity. If the wind shifts, the fire can change direction suddenly. In addition, burning embers, lightning strikes, or accidents can cause multiple outbreaks. The amount and type of fuel, the weather, and the terrain all affect how a fire spreads and how intense it is. And although winds affect how a fire starts, large fires can create their own winds, making the direction and intensity of the fires unpredictable. (Winds of 75 m.p.h. and higher have been recorded during the Southern California fires.)

You can do a lot of things to prepare if a fire is threatening to move in your direction:

◆ Connect your hose to an outside tap. (If you're using an external water supply, set up the water pump.) Wet the shrubs, trees, and lawn 15 to 20 feet all around the house.

◆ Wet down the roof and gutters: even if the roof is fireproof, having it wet will prevent embers from flying off again to find something flammable. (If you have time and there's enough water, you can even set up sprinklers on the roof to keep the roof and surrounding areas moist.)

◆ Clear out any flammable materials near the house, including garden rubbish, patio or pool furniture, awnings, charcoal briquets, and tarps.

◆ Seal the house vents with commercial vent seals. You can use precut plywood seals if you prefer. The goal is to keep sparks and embers from flying into the house where they can start a fire inside.

◆ Keep any pets indoors, preferably in one room. Have pet carriers ready in case you need to evacuate.

◆ Stack fire tools where they're easy to get to.

◆ Don protective clothing: cotton or wool long-sleeved shirts and pants and good shoes. Have gloves and a handkerchief or bandanna for face protection available. Goggles, a face mask, and a tight-fitting cap or hat are also good.

◆ Back your car into the driveway or garage. Leave the car unlocked and the key in the ignition so you can jump into it and go immediately. Keep the windows rolled up so smoke and sparks don't get into the car. (Keep the garage doors closed for the same reason.)

◆ Keep your Go-Paks near the door so you can grab them and head straight for the car.

Because wildfires are so mobile, it's important to keep track of the fire's progress and locations. Escape routes can get blocked off or even burned out. Have several different escape routes planned. Also try

to plan routes that allow you
to evacuate in several different
directions. As the bulletins come
in, keep track of the fire's locations and the road conditions so
you know which of your escape
routes will provide maximum
safety.

Life Preservers

When you're planning
your escape routes, also
plan at least one escape
route by foot.

Feeling the Heat

If you can feel the heat from the fire or see the flames, you're in immediate danger. You need to make some decisions right away about what to
do. When you're advised to evacuate, do so immediately.

Stay, Spray, and Pray

It's unwise to ignore a wildfire evacuation order, but it's possible that
your escape routes have all been blocked off. This may be particularly
likely if you're living in a more remote area where there is only one road.
In this case, you're going to have to tough it out and cross your fingers.

If you're trapped in a potential fire zone, keep your protective clothing
on at all times. Watch for sparks and approaching flames that directly
endanger your house. Keep the house and surrounding vegetation wet.
Contact someone outside the area to let them know that you're not able
to escape and may possibly need emergency evacuation.

One Last Pass for Prevention

If it looks as though you're going to have to evacuate, take a few minutes
for final safety preparations:

- Close all windows and doors and latch or lock them. Also close
  heavy drapes and Venetian blinds.
- Let people outside the area know when you're leaving and where
  you're going.
- Turn off the utilities, particularly the gas. Turn off propane tanks.

- Close the fireplace screen and open the fireplace damper.

- Move flammable furniture and objects away from the windows. If you have lace or other lightweight curtains, take them down: a single spark can set them on fire.

- Move combustible patio furniture inside so it doesn't provide an opportunity for fire to start near your house.

- Wet down the house and landscaping one last time.

Too Close for Comfort

Because wildfires can move as fast as 40 m.p.h. under the right circumstances, evacuation can be a matter of a couple of minutes. Keep listening to bulletins for information about the progress of the fire in your area in case there's advance warning, but don't feel as though you have to wait. If there are flames a quarter mile away or embers dropping in your yard, it's time to go.

A fast-moving wildfire can actually create its own weather front— a firestorm—with hot winds that blow sparks and burning material a considerable distance from the fire. The winds can produce fire whirls, which are basically dust devils made of fire, embers, and superheated air. The greater airflow also increases the amount of oxygen reaching the burning fuel, making the fire even hotter and producing intense radiant heat that can ignite fuel from a distance. Large firestorms can also trigger the formation of pyrocumulus clouds, which can in turn produce lightning and start more fires in the same general area.

Firestorms are extremely dangerous and are frequently responsible for firefighter deaths because of their speed, intensity, and unpredictable behavior.

Storm Warnings

Don't count on being able to outrun a wildfire. Even "slow" fires can easily move 10 to 15 m.p.h. Moreover, gusts of wind can suddenly box you in by causing the fire to leap ahead of you.

Grab-and-Go Evacuation

You should have all your preparations made. The speed with which a wildfire can move into an area means you'll have time to grab your Go-Paks, throw the pets and kids into the car, and leave.

Under unusual circumstances, you may have to evacuate on foot, either straight from your home or if your escape route is blocked. Make sure your car is off the road so it doesn't block anyone else's path, take the Go-Pak from the car and move away from the fire. Try to keep to level ground: if you're uphill from a fire, you can get blasted by smoke, embers, and superheated air if the wind shifts.

If you have no other alternatives, take shelter in a lake, river, or even a swimming pool. A deep ditch may provide enough cover even if the fire sweeps right over you; however, smoke and hot gasses can suffocate you. Keep covered up so that embers or burning material don't land on you. All of these should be considered only as a last resort. Being that close to a wildfire is courting potential disaster.

When Leaving It All Behind

Take with you what is irreplaceable. Ideally, you already have your important papers in a safe-deposit box, but you still have many irreplaceable mementos and personal possessions. Walk through your house and look at your possessions and see what speaks to you as irreplaceable, including:

- Photographs
- Children's toys
- Diplomas
- Trophies and awards
- Family heirlooms
- Mementos and souvenirs
- Jewelry
- Gifts
- Pet toys
- Quilts
- Backup data or hard disks
- Paintings

Lots of things may be irreplaceable, far more than you can carry, but you should make an effort to take things with you when you evacuate. If you lose everything, having as many of these irreplaceable items as you can will have a tremendous impact on how quickly you can get back to a normal life.

The sad thing is that people who live in wildfire zones tend to take everything on the first few evacuations. After that, they take less of what's important, thinking that nothing happened the last three times they evacuated. Finally, they tend to leave everything behind because nothing happened the last half dozen times, and one of the mementoes got broken the time before … and that's when the house burns to the ground. You need to treat each evacuation as real and consider the possibility you may never see your house again.

Related Hazards

The hazards of a wildfire are not limited to the fire itself. Fire is just the beginning. When a wildfire burns through an area, it destroys the ground cover and dries the ground. There is no vegetation left to hold the soil in place. Rain may be temporarily welcome for extinguishing the fire, but it's likely to cause flooding and mudslides.

Wildfires are very destructive to property, but the biggest danger to humans is bad air and smoke. Even healthy people can have respiratory problems from the particulates and gasses in the air. You may have symptoms such as burning or watery eyes, sinus irritation, coughing, or a scratchy throat. More intense or prolonged exposure can also cause chest pain, shortness of breath, and headaches.

People with respiratory or heart problems may show signs of chest pain, wheezing, coughing, fatigue, shortness of breath or other breathing problems, rapid heartbeat, or possibly angina. Children and seniors are more likely to show symptoms from smoke inhalation.

The best solution is to limit your exposure to smoke. Keep your windows and doors closed so that the air inside the house stays as clean as possible. Keep the fresh air intake shut on your air conditioner and make sure the filter's clean to trap particulates. Don't use gas stoves.

Paper dust masks are only effective for dust and other large particulates; they do not filter out smoke. The real danger of smoke is actually from toxic gasses. Carbon monoxide is the most common, but there can be complex and more dangerous gasses created by toxic compounds burning downwind. The safest bet is to minimize your exposure to all smoke from wildfires.

Ashes to Ashes, Dust to Dust

The results of a fire can vary tremendously. You can come through with your house and your community untouched, everything you left behind could be smoke and ashes, or there could be any level of disaster in between. What is left when you come home depends on what burned and where.

A Life in Ashes

Of all the potential disasters and threats to home, property, life, and lifestyle, fire holds the greatest mystery and portent. Fire feels as if it's alive somehow. We talk about fire consuming, dancing, and breathing. You'll often hear firefighters talking about outthinking the fire and how fires want things: to breathe, to consume the building, to skip from treetop to treetop. We talk about fire as if it's a wild animal or foe that's deliberately trying to get us.

As with any time we feel no control over a threat, this means that the power of the threat is magnified. We feel like children against the boogeyman. It's not unusual for people to seek a sense of control through acts of heroism/foolishness. We talk about fighting fires. We're beating our chests at the fire because it's an invader. We'll fight fires with a garden hose or with a bucket brigade. In a small fire that might work, but for a wildfire, we might as well spit at it.

Recovering after a wildfire is different from other disasters. You're usually either not affected by a wildfire at all or else you're very strongly affected. When you come home, if your community's not burned, you unpack your car and return to your daily life. On the other hand, if the fire hit your neighborhood, even if your house isn't burned, someone you know lost their house. They may not come back from this, making this a bit like a death. If your house is burned, you may choose not to rebuild, meaning that you have to relocate and lose your community connections.

A Phoenix Rising

Rebuilding after a fire is a personal choice, whether your house is completely destroyed or only damaged. People who choose to recover and rebuild are those who literally rise from their own ashes. You may be able to do some things to make your house a little more fireproof, such as using a steel or tile roof, concrete siding, and fireproof landscaping, but in the end, there's not much you can do if you're in a wildfire zone. There are many cases where a wildfire doesn't actually get closer than a quarter mile and the superheated air and the intense radiant heat are enough to heat parts of a house to the ignition point.

Being in a potential wildfire zone is a personal choice, just as living near hurricane, quake, tornado, or flood zones are. You're making a choice to be there because it's a hazard you're willing to accept.

The Least You Need to Know

- ◆ Wildfire is not controllable with a garden hose; if wildfire is approaching, the only wise choice is to evacuate.

- ◆ The best protection is to not build in a fire-prone area; the second best protection is to remove as much tinder and flammable debris as possible.

- ◆ As with so many natural disasters, the greatest risks with wildfire extend beyond the fire itself. Flooding, mudslides, and smoke pose risks that extend beyond the boundaries of the fire itself.

- ◆ The nature of fire and its relationship to everyday life gives wildfire a unique emotional impact.

- ◆ Rebuilding after a fire poses unique problems because even surviving structures may have suffered damage invisible to the naked eye due to heat, smoke, and/or water.

Chapter 11

Earthquakes

In This Chapter

- ◆ The causes and types of earthquakes
- ◆ Surviving when the quake hits
- ◆ Evacuating the quake zone and staying safe in the process
- ◆ Determining quake-related hazards: fire, flooding, falling objects, and being trapped
- ◆ Recovering from a quake

Of all the disasters described in this book, earthquakes are probably the least understood and the most feared. We still only have a general idea of what causes earthquakes and how often they may occur or how big they'll be. There's still no reliable way to predict them, which makes it all the more important to be ready for when they do strike.

In this chapter, we look at what causes earthquakes and where they're most likely to occur, specific things you can do to prepare your house, what to do during a quake, how to deal with the collateral damage caused by a quake, what to do if you're trapped, and getting back to normal.

What's an Earthquake?

The Earth is not solid. Large sections called tectonic plates move slowly around on the surface. As these plates move, they rub up against each other and can build up pressure along the edges. Tectonic plates typically move between one or two inches a year. Even though some plates move as fast as 5 inches a year, this still only works out to 1 mile every 13,000 years or so.

In the best circumstances, these plates slide fairly freely over or under each other. However, plates can stick or press against each other and pressure then starts building up, just a little at a time, but these are continent-size plates. The edge where the tectonic plates rub against each other is known as a fault line, and this is where earthquakes tend to happen. When the pressure is enough to overcome the friction, the pressure releases suddenly at the fault line and there are tremors on the surface: an earthquake.

Bump and Grind

There are two basic kinds of earthquakes: slip quakes and thrust quakes. Slip quakes happen when two plates are pressing against each other and the pressure is released by a sideways motion.

In a slip quake, the ground over the intersection of the plates is disrupted and the top layers fall into the void. Slip quakes look similar to the classic Hollywood disaster-film earthquake, where a large rift opens in the ground. Damage such as long rips in roadways and buildings that have been split into two pieces several feet apart is typical.

Mountains from Molehills

Thrust quakes happen when the pressure between two plates is released when one plate pushes up (and possibly over) the other plate. (Think Hollywood sci-fi films from the 1950s here.) The ground over the colliding plates is pushed upward as the plate edges are forced upward. This sudden upward motion can cause tsunamis if they happen underwater. On land, thrust quakes can push things over. Thrust quakes are ultimately responsible for new geological landmarks such as mountain ranges.

Liquid Assets

An interesting (and potentially dangerous) thing happens to the ground when it gets shaken: it starts to act as a thick liquid. Liquefaction is very much like quicksand but on a much larger scale. The liquefied soil transmits the vibrations from the quake to whatever is sitting on it, magnifying the potential damage from the waves tremendously.

The real danger of liquefaction is that whatever is sitting on top of the ground can sink, sometimes suddenly and dramatically. Things don't always sink evenly, so houses and buildings can have one side that's now several feet lower than the other.

How much liquefaction there may be depends on the ground itself. Solid ground with a bedrock base probably isn't going to liquefy at all, but something that's fill dirt and rubble 30 feet down can become pudding in a matter of seconds.

Measuring Quakes

Earthquakes have traditionally been measured by the Richter scale, developed in 1935 by Charles Richter. This scale is a logarithmic scale, so that, for example, a 7.0 quake is 10 times as powerful as a 6.0 quake.

The smallest earthquakes you're likely to notice are 3.0. Quakes of intensity 4.0 rattle things, but cause no damage. Quakes of intensity 5.0 might cause some amount of building damage, but rarely loss of life. Quakes of intensity 6.0 cause major damage and substantial loss of life when they hit anywhere near a population center. And anything 7.0 or higher means massive destruction and death.

The Richter scale doesn't accurately measure the size of quakes higher than 8.0, so the moment magnitude scale was introduced in 1979 as a replacement for the Richter scale. This is now the official scale used by the USGS and seismologists. The media and even public information officers from the USGS say "Richter scale," but are actually reporting the moment magnitude scale.

We're Gonna Rock This Town

There are lots of areas known for seismic activity in the United States, most of which are on the Pacific Rim: California, Oregon, Washington, and Alaska. However, all states have some seismic activity periodically. For example, the New Madrid quakes of 1811 and 1812, estimated to be around magnitude 8.1, happened in Missouri.

Alaska has the most earthquakes of any state, followed by California. Florida is the most seismically inactive state, but Florida has hurricanes, so take your pick of the type of disaster you prefer.

Earthquake risk is like everything else in real estate: it's all about location, location, location. Earthquakes occur along fault lines, so prime earthquake regions are well documented and roughly predictable. If you build something right on a fault line or in an earthquake zone, you're accepting that sooner or later you're going to be taking damage from a quake.

Liquefaction is more of a concern if you're building something in an earthquake zone, but it's also good to know where major liquefaction zones in your city are, too. In Seattle, for example, there are large areas of regrade that are much less stable than some of the surrounding land. If you know you're in a potentially hazardous area, you may be able to make an informed choice about getting out of the area as quickly as possible after the initial quake in case of aftershocks.

Batten Down the Hatches

In the last few decades, building codes in most areas likely to have earthquakes have been updated to include substantial earthquake safety measures. But older houses may need retrofitting, particularly if there hasn't been a major or lengthy earthquake since the house was built. You may need to do some structural retrofitting to make your house safe.

Bolts, Straps, and Retrofits

One of the most common problems for older houses is that the structure jumps or slides sideways and skates off its foundation. You can have the house attached with special connectors to the foundation to prevent this.

The foundation itself may also need reinforcing: for example, an old pillar and post foundation could fall sideways in an extended earthquake that got the house swaying back and forth. Additional supports and braces may be required. Chimneys, particularly those that are much higher than the roofline, can crumble and fall through the roof. You can add reinforcing and support, or you may even want to replace the entire chimney with lighter, stronger insulated pipe.

Foundation and structural work is expensive, but if you're in an earthquake zone, it could be worth it. (Sometimes it can result in a reduction of earthquake insurance rates, too.) If your house falls off the foundation, it's probably going to be a write-off. Remember, though, that retrofitting and even completely earthquake-proof new construction isn't proof against any quake. It just means that you're most likely to survive an earthquake with as little structural damage as practical.

Of more immediate concern is what's happening inside. Quakes big enough to shake your house off its foundation are thankfully rare. Quakes that can knock things off shelves are far more common, and it's this type of damage that you have to prepare for. A 5.4 quake can, for example, do any or all the following:

- Open cupboard doors and throw coffee cups or plates on the floor.
- Throw books off shelves.
- Knock paintings off the wall.
- Tilt entire bookshelves over.
- Move the refrigerator out 3 feet.
- Drop a ceiling-mounted light fixture to the floor.
- Shake a TV out of an entertainment center or wall niche.
- Swing a hanging plant or wind chimes into a window.

Look around your house and see where there are large or heavy objects that can fall over (such as bookshelves) or fall down (such as a painting).

The first thing to do is put earthquake strapping on anything that can fall over. For bookcases, grandfather clocks, and refrigerators, this is usually nothing more than a piece of heavy woven fabric band and a couple of screws to anchor it in the back of the object and the wall. You don't need something that can bear the weight of the object falling forward; you just need enough to stop it when it rocks out a few inches.

Statues, souvenirs, and other knickknacks on shelves and mantles can be held in place with museum putty or gel. These are neutral adhesives that hold items in place so they can't be shaken loose. Consider adding a monofilament line an inch or two, held in place with thumbtacks or a little tape, above each shelf to keep objects from bouncing off the edge.

Paintings and hanging mirrors can be secured with heavier hooks, a second wire (as a backup in case the first one breaks free when shaking), or corner brackets that anchor the frame securely and prevent it from swinging.

Other things to make your house a little safer:

- Move heavier items to lower shelves.
- Install latches on cupboards.
- Keep cleaning supplies, flammables, and other chemicals where they can't fall and break open.
- Check overhead fixtures and hangers.
- Add a monofilament guy line to hanging plants or wind chimes so they can't swing into anything breakable.

Quakeproofing Your Utilities

Your utilities need to be quakeproofed, too. Start by checking your wiring, gas lines, and plumbing. If you have frayed lines or weak joints, replace them. The jerking motion of a quake can tear electrical lines and rupture gas lines, causing fires, and crack water lines, causing flooding.

If you use gas, check with your gas company about getting an earthquake shutoff valve. The most common of these is basically a valve with a large ball sitting on a post inside it. If the ground shakes, the ball is knocked off the post into the pipe below, blocking the gas flow. There are also gas and water "leak cut-off" valves that detect broken pipes based on pressure changes or on water accumulation. There are even seismically activated electricity shutoff valves, too.

There are special strapping kits for your hot water heater that bolt into the wall. They take about an hour to install. Even if the chances of an earthquake in your area are low, put one in just in case. You should also strap any other gas appliances in place so they don't "walk" out from the wall and break open the gas connection.

Little or No Warning

Unlike most other disasters, earthquakes usually strike with no warning whatsoever. Unfortunately, there is little warning for earthquakes even in the most sophisticated of monitoring environments. Accurately predicting when failure point stresses have been reached is beyond the ability of current technology. You have to make plans to respond rapidly to ensure your safety and the safety of your home.

Personal safety is mostly ensuring safety from falling objects during the quake itself. As soon as possible after the shaking stops, you need to evacuate quake-damaged structures. The worst place to be when an aftershock occurs is in a damaged building. Ensuring the safety of your home means having automatic utility cut-off systems in place to prevent collateral damage from fire or flood and having furniture, appliances, and other heavy objects suitably anchored so they don't fall.

Seismic Activity or Lack Thereof

If you live on the Pacific Rim, you're used to the idea of earthquake preparedness. But quakes, even big quakes, can happen literally anywhere at any time. Fortunately, they're few and far between but they happen without any warning. Be prepared for when the inevitable quake strikes.

High-Tech Home Tech

Many new building materials are designed for earthquake safety. Although these are more applicable to new construction, some of them are designed to be put on as part of a quakeproofing remodel or retrofit.

One of the simplest techniques is to install windows that have the glass panes not touching the corners of the frame. This lets the glass move slightly in the frame during a quake rather than crack and then shatter because it's pushed into the glass frame. Rounding and polishing the corners of the panes makes the windows significantly less likely to break.

Another quake retrofit is to add security films to windows. These won't prevent the glass from shattering, but will prevent shards of glass from covering an area.

Tall buildings may have inertial damping shock absorbers as part of their foundation. These are literally giant shock absorbers that absorb the sudden shock of a quake. Another trick is to have a large tank of water with a heavy pendulum near the top of the building. When the bottom shifts, the top doesn't respond quickly because of the inertia of the tank and pendulum. The building will shake as a result of the vibration, but it doesn't move as much or as quickly. Instead, the pendulum absorbs a large portion of the momentum, keeping the building safe.

Evacuation Issues

Where you are the moment a quake strikes will make a big difference to how you respond, but in all cases you need to make sure that you avoid getting hit by falling debris or being trapped under rubble.

Finding a Safe Haven

If you're indoors, the first thing *not* to do is go stand under a doorway! Doorways aren't any stronger than the rest of the house. Worse, you can get smacked by the door if it swings back and forth and you still can get hit by falling debris. It's better to leap under a desk or a sturdy table—something that will provide you with space even if the ceiling falls on it—and stay there until the shaking stops.

If you're in bed, stay there. Cover your head with pillows and use the blankets to protect your body. Don't hide under the bed: if the bed collapses, you'll be crushed.

If nothing else, huddle in an inside corner of the building where you're not likely to be hit by anything falling. Keep your head and face covered.

The Security of the Street

If you're driving, treat a quake like a blowout: brake as quickly and safely as possible, and pull over to the side of the road. Be aware that everyone else is trying to do the same thing, weaving on the road from the quake, or both, so it may be a challenge. Stay in your car until the shaking stops. Driving after the quake ends may be dangerous because of rips in the asphalt, broken or weakened bridges and overpasses, and the possibility of aftershocks.

If you're walking down the street, duck into a sheltered doorway or archway, but keep an eye on what you're standing under: loose façades, signs, and even plate-glass windows can break and fall on you.

Speaking of windows, if you're surrounded by high-rise buildings, you may have a problem. Many tall buildings are designed to twist when hit with a large quake. The buildings don't fall over, but the panes of glass (some of which weigh as much as 300 pounds apiece) can pop out of their frames and fall to the street below. In this case, get under cover as fast as possible, preferably behind something that will block glass shards. Similarly, get away from utility poles and streetlights, which can fall over.

Another hazard during an earthquake is being surrounded by people. Whether a sporting event, a movie, or a mall, when an earthquake strikes, large numbers of people are going to run toward a limited number of exits. As these exits become clogged with a human cork of bodies, fear rises and panic ensues. An orderly crowd becomes a raging mob. The crush of human bodies causes people to be injured and even suffocated.

The crowd will be going the direction you want, but you're likely to meet up with a cork of bodies and you may be stuck inside where the hazard is. When you're headed to the doors with the crowd, you're the one that's going to get shoved out the doors.

Tips for surviving a stampede:

◆ Don't freeze up. Keep moving and act decisively.

◆ Don't panic. Keep as calm as possible and don't push against the people in front of you. If they fall down, you'll fall on top of them.

◆ Stay in the crowd rather than against a wall, where you can get crushed. The biggest danger in a mob is being crushed and suffocated. It only takes a minute or two to get suffocated to unconsciousness.

◆ Keep your hands near your shoulders and your elbows up to create the largest space around you. Your hands are at the ready and your elbows will hook and catch on people next to you if you start to fall.

◆ If you have possessions, carry them above your head. It'll make you that much thinner to the crowd.

◆ Don't try to buck the crowd. You won't win and you could be trampled.

◆ If you drop a possession, leave it. Even reaching down for it puts you at risk for being knocked over.

◆ If you're with someone, don't let go of their hand unless you are drawing them into a more dangerous situation, such as a crush of people.

◆ If you have small children, put them on your shoulders and piggyback them out of there. For bigger kids, grab their shirt collars or their belts. (They can't fall if you're hanging on to their belts.) Don't do this with a teen child or adult who's your size. In fact, if they're bigger than you, have them hang on to your belt for the same reason.

◆ If you do fall, get up as fast as possible. You can reach for someone or grab someone's clothes to pull yourself upright. Reach for waistbands: if you start pulling their pants off, they'll pull you up. Don't grab arms and legs; they'll push you off. Failing that, crawl in the same direction as people are moving and keep looking for opportunities to get to your feet.

Related Hazards

It's 30 seconds later and the ground has stopped shaking. You weren't beaned by anything or buried in wreckage. So far, so good. Now you have a different set of worries.

The biggest danger after an earthquake is fire: broken gas lines, sparks from frayed electrical lines, and even things that have fallen on hot stoves and ignited. You'll know if something's on fire on the stove right away. Grab a fire extinguisher and put it out quickly as best you can, then get out!

The most likely flooding you're going to encounter after a quake is locally from a broken water main, but it's just barely possible that a reservoir or even a nearby dam could rupture. Nevertheless, your greatest water-related danger is the old rule of power lines and water not mixing. Stay away from downed lines of any kind and remember that electricity can travel a long way through standing water.

The initial quake will tend to knock things off shelves and off buildings, but objects can also be loosened or moved right to an edge. All they need is one little nudge and over they'll go. It doesn't even have to be an aftershock: the vibration from one good door slam might be all something needs to fall down. Be aware of what you're walking under or near.

Getting Trapped and Getting Free

Most earthquakes don't cause enough damage to collapse structures, but it only takes one building or freeway falling on you to spoil your whole day. Fortunately, you can increase your chances of surviving three stories of building on top of you.

If you're hiding under a table and the roof falls in, you can be safe for the moment, but trapped. The first thing to remember is never light a match. If there's a gas leak, you can ignite the gas and be unable to escape. You can use your cell phone for enough light to see where you are.

When you're trapped, you can live for weeks without food, but only days without water, and a few minutes without air. (Don't freak out about the air when you're trapped: if you're alive to notice it, you're probably okay.) Breathe through your nose; you'll conserve water.

Don't scream aimlessly: it wastes energy, air, and water. Be quiet and still, listening for people trying to find you. Reach around slowly and gently (being careful of sharp objects) to find something to bang with. Metal on metal's best, but almost anything will do. Rescuers will be listening for this kind of noise. Get these people to come find you.

Depending on the circumstances, rescuers may not be there for hours or even a day. Sleep when you can. It'll be easier to do than you think: if the boredom doesn't get you, the stress and exhaustion will. Sleep's actually the best thing to do right now, as it conserves energy and water. Don't worry about sleeping through the rescuers. They're going to make a lot of noise and your brain's going to be listening for them. When you hear people outside, that's the time to yell and bang things as loudly as you can. Pause every so often to listen, then yell and bang some more.

You may have water dripping into your space. Don't drink it; in fact, try to avoid touching it. You have no idea where it comes from. It could be straight sewage. If you're trapped for a while, you could actually die from dysentery long before lack of water or starvation would've gotten you.

There are two problems with trying to dig your way out. The first is figuring out which way is up so you don't try digging to China. Seeing which way the dust falls or a handful of dirt or debris will give you a pretty clear idea of "down." The second problem is much bigger: determining the best way to dig. Sure, if you kept digging up, you'd eventually get through the last layer of whatever's on top of you, but there might only be a foot of debris to your left. Worse, it's possible that you may be digging up through 20 feet of rubble, which will then fall on you. By digging, you can take a stable entrapment and potentially turn it into something unstable.

As a rule, it's probably a good idea to stay where you are. However, if you can see daylight in any direction, dig toward that. As you move debris, listen for the sounds of shifting or falling debris. Be ready to dive back under your table if it sounds as though things are about to fall.

It is pretty rare for people to extricate themselves. It's usually the bystanders who do the rescuing at first, followed by urban search-and-rescue teams.

If you're trapped in a car, the car has probably formed a bubble of safety around you. As long as there's no immediate danger, it's probably best to stay there instead of disturbing the equilibrium. On the other hand, if you're in what appears to be an unstable situation or there's a danger of your car filling with water, get out immediately.

After the Shock

The most important thing to do after an earthquake is to evacuate the building until you know it is safe to return. Ideally, gas, water, and electricity will be cut off immediately. A good alternative is that the utility cut-offs are on the outside of the building where they can be safely reached after you evacuate the building.

Although it's understandable that you want to return home and look for gas and water leaks and electrical hazards, it's just too dangerous. If there's any indication of quake damage, do not re-enter a quake-damaged structure for any reason until professionals have determined that it is safe to do so!

Aftershocks

Earthquakes may come suddenly, but they can take a while to leave. Frequently, there will be aftershocks: smaller earthquakes that happen anytime in the next few days. Apart from the surge of adrenaline these cause, there's a very real danger that any aftershocks will be just enough to complete damage that was started by the previous quake.

Building Safety

Apart from not going into any structure that shows any signs of damage, earthquakes cause some subtle types of damage to buildings that don't show up right away. Listen for odd noises from the house that suggest settling. Be on the lookout for signs of shifting or cracking in the walls or foundation. One sign of settling is doors, windows, or built-in drawers that start binding or jamming.

Settling may not indicate damage to your house directly caused by the quake, but may reflect a change to the soil conditions under the house. For example, a water or sewer main may have cracked nearby and the additional liquid is now causing slipping in the soil. As the soil loosens, there may be collateral damage to the house as it settles. Be sure to have any settling after a quake checked out promptly to prevent further damage.

Windows are a regular casualty of earthquakes. Expect the possibility of broken glass almost anywhere and don't go barefoot.

Restoring Utilities

If you have an earthquake shutoff valve for your utilities, you'll need to wait for the various utility companies to reset the valves. On the other hand, if you just turned your utilities off at the mains manually, you can turn them back on yourself. (Check with your gas company to see if they have a training program about turning your gas back on.)

Power and water can be turned back on, but it's a good idea to inspect what you can of the wiring and plumbing. Don't turn things back on and then go to bed. Turn them on so you can check and then check again in a few hours. If there are electrical or plumbing problems, they may take a while to manifest.

Drinking Water Safety

Just because the water may be flowing doesn't mean it's safe for drinking. Earthquakes can cause cracks in the water mains up the line from you that allow groundwater and contaminants to seep in. Well water will almost certainly be murky because it's gotten sloshed around. It's even possible that septic tanks and sewage pipes have cracked and leaked, causing invisible bacterial contamination.

Until you get the all-clear from the local water authority, treat water from the tap as if it's contaminated. Boil or sterilize water before drinking or cooking with it. If you're on a well service, pump the well out and then test the water.

The Least You Need to Know

- ◆ Earthquakes occur with little warning. Plan ahead for your safety and that of your family.

- ◆ Don't be lulled into complacency by a lack of earthquakes; if you live in an earthquake zone, another quake will come.

- ◆ The safest place in an earthquake is in an open field or park, far from buildings or other items that can fall.

- ◆ If you should be trapped, conserve energy and water until you hear people, then yell for help or bang objects.

- ◆ After the quake, safe shelter and clean water are the most important things to have.

Chapter

Blizzards

In This Chapter

- ◆ What a blizzard is and the real hazards of snow and ice
- ◆ Keeping you and your house warm
- ◆ Getting ready when a blizzard's on the way
- ◆ Removing snow safely and without risking a heart attack
- ◆ Thawing out in general

If you grew up in the north, you remember the joy of "snow days," those wondrous days when school was closed and there was nothing to do but ride sleds and build snowmen. Unfortunately, there is an ugly side to snow and the winter blizzards that create snow days. When winter winds howl and grey/black clouds unleash a wet, cold blanket over your city, snow days can quickly become a fight for survival.

In this chapter, we take you through the winter wonderland of blizzards and cold-weather disasters. We help you prepare yourself and your home to survive snow, wind, and ice whether you choose to shelter in place at home or evacuate to Florida. We discuss not only preparations, but what to do if you find yourself suddenly cold and alone.

What's a Blizzard?

A blizzard is a storm with a lot more snow than usual. This means that a blizzard can happen anywhere it can snow. The amount of snow that makes a blizzard is something else, however. What defines a blizzard is how much snow you're used to. Seattle got 12 inches of snow on December 26, 1996, and the city shut down for almost a week. On the other hand, parts of Minnesota, Michigan, Wisconsin, Vermont, and New York would merely think of 12 inches as a heavy accumulation of snow.

Blizzards are usually associated with other weather conditions that cause problems. Usually, there's wind: it's not just falling snow, it's blowing, falling snow. This can result in white-outs. The snowpack is frequently different: wetter, heavier, and more prone to icing. So a blizzard isn't just more snow than usually. It's frequently lots more snow coming down in a different way with a very different characteristic.

Snow

The two big issues in a blizzard, unsurprisingly, are snow and ice. When you have a generally flat surface, as many of the Plains states, you have blowing snow and lots of it. The winds pick up the snow and blow it for miles until it hits something to stack up against. Even if the snowfall was less than a foot, houses, barns, cars, and trees can all be quickly buried under drifts of snow.

Any large accumulation of snow in the mountains increases the potential for avalanches. However, a rapid fall of snow can form a layer prone to slipping (particularly if the layer underneath is icy), making an avalanche extremely likely.

Avalanches are only fun at a distance. When the top layer of snow starts moving, it acts a lot like a landslide or a mudflow. Large blocks of snow and ice can be bounced along and trees and boulders can be picked up and carried by an avalanche.

Avalanches are deceptively fast, too. A small, slow avalanche that's mostly large blocks of snow moves at 20 m.p.h. These avalanches tend to be very destructive, though, as the snow is wet and very dense. Larger avalanches move more than 100 m.p.h., and a light powdery avalanche (similar to the ones you see in movies) can move in excess of 200 m.p.h.

Avalanches don't just bury people; they can cut off whole areas. Colorado, for example, has avalanches that isolate resorts for days at a time.

Ice

Ice can form at high altitudes and fall as hail, but it more often forms when snow or wet snow blows on something and then freezes. The real danger of ice isn't from icy roads (though that's bad enough). Ice is heavy and it can accumulate on things that don't normally get a lot of weight. Unlike snow, which is a problem for flat surfaces, ice can cling to tree limbs, power lines, and the like, and then rip them loose.

Besides weight, ice adds wind resistance, so trees that normally withstand the wind now have ice "sails" on them. Even the limbs that weren't heavy enough with ice to have been ripped out can be torn out by the wind and fall on things and people below. When power lines accumulate ice, it can break the lines between each and every pair of utility poles for tens and even hundreds of miles of cable. Splicing is impossible for damage such as this; it's necessary to run new cable. Prince Edward Island lost literally all its power lines this way in 2008. It took several weeks to restore the power lines, and the lack of power added to the dangers and risks to the population.

Objects on the ground, such as cars, can be slowly encased by ice up to several inches thick when there are alternating freezes and thaws. The surface of the ice gets tacky or moist, more snow falls, then freezes in place.

Wrapping Up in Cold Weather

Blizzards can happen everywhere and anywhere it snows, although some areas are more likely to experience blizzards than others. The likeliest places for blizzards in the United States are the Great Plains, the Northeast, and the Northern Rockies and Cascades.

Insulation

Besides staying warm, the best thing you can do to stave off the cold is to wrap up. This is true whether it's you, your pets, or your house.

Insulation works by trapping air in dead spaces between the cold and something warmer. Air is an exceptionally poor conductor of heat. When you insulate a house, it's not a one-time event. Insulation loses effectiveness as it ages. As it slowly packs down, it loses air.

Animals and even insects degrade the insulation, too: bits of spider web and damage from rats or sparrows can compact insulation and leave channels that air is freer to travel through. Most damaging of all is if insulation gets wet. This can trap moisture in your walls and cause dry rot as well as cause the insulation to pack down to the point of uselessness.

The location of the insulation determines the type and value of the insulation. Rock wool and shredded paper insulations can be sprayed on in blankets, making them good for attics. Fiberglass rolls and batts and mylar ply are very good for wall insulation. Injected foam is effective but it's now only used in concrete blocks because it breathes toxic chemicals.

Windows

Insulation's effectiveness is measured by its R-value: the higher the R-value, the better the insulation. Windows and doors are now routinely measured with their R-value.

The things that determine a window's R-value include ...

- How many panes of glass/layers of air there are.
- How much light is transmitted (and how much heat).
- What the window frames are made of.

Older houses frequently have single-pane windows, which are notorious for wasting heat. Most municipalities offer zero-interest or low-interest loans for replacing single-pane windows with double-pane insulated windows.

You'll often see listings that mention "low-E" glass. The glass in the window itself is coated to block infrared radiation, so heat inside stays inside. In addition, multiple panes of glass are used to create dead layers of air, which are often sealed and filled with nitrogen for greater insulating value. Even storm windows provide a small amount of insulation because the shutters or covers add a layer of relatively nonmoving air. It's not great, but it's better than nothing.

Along the same lines, you can apply window film to existing windows. The film traps a small layer of warm air between the film and the glass. Most window films also have some value for cutting down on infrared losses.

Window frames can be wood (an excellent insulator) or plastic or aluminum, which is then usually placed next to gaskets and insulation. But no matter how good the window, it'll have a lower R-value and leak more heat than the surrounding wall.

Doors can have very high R-values depending on what they're made of and how well they're weather-stripped. Even something as simple as putting a tightly fitting block of foam in an embedded window can make a substantial difference.

Fuel

With the house insulated, you need to heat it. The question is which and how many types of fuel you can burn to stay warm. Most houses have a single primary fuel source—coal, oil, natural gas, electricity— with one or more secondary heat sources. For example, a house with a natural gas furnace may also have some baseboards. In addition, there are other heat sources you may be able to use. For example, if you have an air-conditioning heat pump, you can run it backward in emergency mode to provide backup to the primary heat source.

Similarly, a home with a wood-burning fireplace can be used to provide emergency heat. A wood-burning stove is an excellent source of radiant heat. In a pinch when the power goes out, you can do basic cooking on the surface, even if it's only making a large pot of stew.

The major danger when you have the house sealed up tight to keep the heat in is that you can get a buildup of carbon monoxide (CO). CO is an odorless, colorless, toxic gas that is the leading cause of fatal poisoning in the United States. CO forms whenever there's incomplete burning and is present in some quantities in all exhaust and smoke.

In any situation where there is even the slightest chance of combustion gasses in a closed space, you should have CO detectors in the house.

Actually, most jurisdictions that allow for an actively burning heating system of any kind require CO detectors. Many of those also require auto-cutoff switches for furnaces, and all of them require flue caps and flue screens to keep the flue from being blocked by snow and the CO backing up down the chimney and into the house. Flue caps are fairly effective, but if the flue cap is 18 inches from the roof, and there's 30 inches of snow on the roof, you might still have a problem.

If you live in a place that has a relatively short winter, you may be interested in catalytic converter or "flue-less" fireplaces that burn natural gas or a gel alcohol. These units have no external flue. There's a catalytic converter inside the natural gas units that converts the CO to carbon dioxide, which is then vented into the room. The gel alcohol versions may or may not have a catalytic converter. Alcohol fuels burn cleanly, but can be dangerous, too: if the burn's not pure for any reason, you end up with a dose of CO in the room.

Camp heaters come in two forms: inside and outside. The outside versions (patio heaters and the like) are designed to heat an outside area. The indoor models burn the same fuels but they also have a catalytic converter so they're usable inside a tent or closed structure. These units require scrupulous maintenance so they operate cleanly.

Clothing

Insulating your home is one thing, but how about yourself? The old solution of adding layers traps more air between you and the rest of the world. A coat has fabric, padding, even down, all of which trap air, keeping your body's heat from escaping. Insulated clothing traps warm, dry, noncirculating air between you and the rest of the world. You're walking around in a small balloon of warm air heated by your body.

What if you get trapped outside and need extra insulation? One of the things discovered over a century ago by street people is that if you crumple newspaper between your top layer and the next layer, you can add dead air space. You'll crinkle as you walk and you will lose some R-value as the paper crushes down, but the crumple and twist of newspaper can provide enough insulation to save your life and keep you quite warm.

When you're using paper as insulation, add layers of uncrumpled paper in your shoes. Paper is almost as poor a conductor of heat as air. Also stuff paper in your stocking cap. In fact, on a baby or small child, you can save 30 percent of the heat in their body with good head insulation.

You don't have to use newspaper specifically. You can do this with anything that holds air, including dry sponges, popcorn, or packing peanuts, but newspaper is readily available.

Weatherize the Car

You can do several things to help your car survive the winter. First, protect it from salt. Use sealant underneath. This is not even an option anymore in some parts of the country: areas prone to salt corrosion get sealant automatically.

You also need to make sure your battery is capable of delivering enough juice—look for "cold cranking amps"—and that you have the correct ratio of water to antifreeze. It's a common misconception to fill your radiator with pure antifreeze as opposed to a mix. In fact, antifreeze may have a higher freezing temp than water/antifreeze. Antifreeze may also have a lower boiling point.

Finally, make sure you have higher air pressures in your tires. As the temperature drops, the pressure drops in the tires, so you're riding on underinflated tires which are more inclined to slide. Conversely, you need to reduce the pressure as the temperatures warm in spring so you don't hydroplane or blow out tires.

Winterizing your car isn't just about the mechanics of the vehicle. You also need something for the passengers:

◆ A mylar space blanket or a regular blanket, so you can stay warm

◆ Lighted warning devices that are independent of the vehicle's battery (flares work, but you have to keep lighting them)

◆ A working cell phone, OnStar, or other system for calling for help

◆ A small shovel and a sandbag, which adds weight for a rear-wheel-drive and adds traction in slippery situations

Wherever you go, it's critically important to keep your Go-Pak with you.

Snow Removal Equipment

There are two tools for removing snow: a snow shovel and a snow blower. A snow shovel is really a hand plow. The majority of back injuries are because people lift the shovel with a load of snow on it, which is an equivalent force of 200 pounds at your hands. You should instead push the snow off to the side. The lifting action is only designed to flip the snow when you get it where you want it to go. There's a thick edge to the blade so when you scrape it on the ground, you don't wear it out. You can also use it to flip ice off the ground. If you use a snow shovel correctly—like a plow—there's no chance of back injury because there's no twisting and lifting.

There are worse things than throwing out your back. If you use the snow shovel like a shovel, lifting snow and throwing it, it's the equivalent of a cardiac stress test. Even worse, the stress on your system goes from 0 to 100 instantly. (A cardiac stress test builds up slowly.) You're also likely to be out shoveling for an hour or so, as opposed to the 9 to 16 minutes that a cardiac stress test lasts. With all this uncommon strain, it's not at all uncommon for people with perfectly good hearts to suffer angina and the like from this.

The other tool is a snow blower. A snow blower is really a snow auger. There's nothing that really "blows" the snow; it just moves the snow into a chute and pops it out the other end. The danger with a snow blower is that anything that gets caught will move through, which takes you to the other name for an auger: meat grinder.

For safety, never wear loose clothing, because at the speed the auger is moving, if you catch the clothing, it'll take the limb with it. The only place to be when the engine is running is behind the blower at the controls. If there's any reason to check anything, turn the engine off completely and take the key out of the ignition or engage the engine safety lock.

Most injuries with snow blowers are because something gets in there that shouldn't, such as a hose, an electrical cord, or a piece of tree limb. Another cause of accidents is when the operator goes around to the front of the thing while it's running and either gets things thrown in his face or gets clothing caught.

You also need to be careful about applying salt and sand to public side-walks. As snow melts, it dilutes the salt, which can then refreeze. You may end up liable if someone then slipped on the refrozen ice. In addition, the runoff from salt and sand can be your responsibility in some municipalities. If you kill the grass and flowers on a nearby median because you salted your sidewalk, the city may bill you to replant them. Check with your city or county government to find out more about any restrictions or guidelines for using salt or sand in your area.

Winter Weather Warnings

As with hurricanes, winter weather tends to give a lot of advance warning. Although a snowstorm can turn rapidly to an ice storm with little opportunity to prepare heavily, you'll usually have a few hours to get ready. However, if you've been diligent about making plans, even a few hours of advance notice should provide enough warning to put final preparations in place, or even to evacuate if necessary.

In most cases, people shelter in place rather than evacuate in a blizzard. Blizzards cover very large areas and most people fail to estimate how bad the storm is. In part, this is because areas that are prone to blizzards get snow regularly and there's a feeling of "It's just more snow, so how bad can this be?"

If you're sheltering in place, expect to lose electricity.

Evacuation shelters are frequently in the area being hit by a blizzard, so evacuation may not be a viable option. Many people in blizzard areas tend to "self-shelter" by going outside the area. Florida's nice that time of year.

If you do need to leave your own house, most shelters in a community have their own heating and power. However, if the shelter isn't providing heat, you should prepare to be bundled up.

Evacuation Issues

Evacuation before a blizzard can be difficult. It's rare that a blizzard is the first snow of the season, so the roads are already going to be snowy or icy. Because blizzards cover such a large area, it may be difficult to make it to an area that is relatively snow-free.

Furthermore, as the blizzard is rolling in, you can have white-out conditions that prevent you from driving, and that almost guarantees accidents and road blockages. Evacuating versus sheltering in place should be evaluated for the relative risks. If there's a good chance you could get stranded on the road somewhere, it may be safest to shelter in place.

Sharing Houses

House-sharing combines some of the aspects of evacuating and of sheltering in place. It also provides a certain degree of esprit de corps. When you shelter with a favorite neighbor or friend, you have more fuel because you're pooling resources. This is mostly a psychological benefit: it's only slightly better than staying in your own house, but because blizzards are often near the holidays, it makes for a more family feel. Sheltering together can turn what could be a lonely time huddled in the dark into something a bit more festive.

Shelters

Even though you may be stuck in a blizzard area, you may still want to shelter locally. Emergency shelters provide a number of advantages that sheltering in place may not. First, shelters will usually have their own generators and heating. Even though you're likely to lose power in your own home, a shelter will likely have electricity and heat. There will also be the advantages of food, community, and some amount of emergency medical care available.

Waiting Too Long

Blizzards are a bad thing to be outside in, but you may not have a choice. If the blizzard is hitting a large area or develops suddenly and you're traveling in your car, you may get caught outside. In such a case, find shelter as soon as possible. Keep moving at a safe speed, even if that speed is only 5 miles an hour. Avoid stopping on the side of the road. If visibility is low, do what you can to make sure that you're not going to get hit by another car.

Above all, don't get out of the car. You can easily get hit by another car driving by. Worse, if there's a white-out, if you slip and fall, you may literally be unable to see your car and you'll end up wandering off in the blizzard trying to find it again.

Carry your Go-Pak in your car, not your trunk. If your car goes into a ditch and you get trapped inside, you won't be able to get to your trunk. Your Go-Pak might as well be at home in a closet at that point. An effective survival trick is to carry candles and matches in your car's glove box. If you're trapped, you can crack a window slightly for air and burn a candle. The heat generated from a single 1-inch candle in a car isn't enough to keep you warm and toasty, but it's enough to keep you from freezing.

If you're caught outdoors and there's no other shelter nearby, look for conifers with low-hanging branches. They'll provide enough of a roof that a heavy snowfall won't cover you and there's enough shelter that you may be able to stave off the worst of the cold.

Related Hazards

It's not just snow and ice that are the hazards. Blizzards come with their own related problems.

Frozen Pipes

A common problem during blizzards is frozen pipes. Water expands when it freezes. A closed pipe in a closed system will rupture if the water freezes. As the pipe defrosts, the water will expand again.

Keep a trickle of water going through your pipes at all times to prevent the pipes from freezing. This is most easily done by finding the valve that's farthest from the incoming flow in your home and leaving a small trickle on.

Many outdoor faucets and pipes in colder areas will have a secondary shutoff valve. When cold temperatures make it likely that outdoor pipes may freeze, you can shut off the water to the outside pipes, then open the faucets so that the water drains from those sections of pipe. The faucets may get way below freezing, but there'll be nothing in them to freeze.

For older homes, you may want to look at pipe warmers: electrical plug-in clamps that have a small heating element that heats the pipe. They're usually outside the house, and in some cases they're placed inside the house and are used to defrost. These are useful if, for example, you have lead sewer outflow pipes. Again, they're very effective, but they cost money. Wrapping a pipe warmer with insulation is a good idea.

Frostbite and Cold Injuries

There are two types of cold injuries: frostbite and hypothermia. Frostbite is the loss of circulation because of the freezing of tissue. Because water expands when it freezes, the cells break open (just as if they're small water pipes). The solution to frostbite is very slow rewarming to 95°F, slightly below body heat. This preserves as much tissue as possible and avoids causing further damage. (One traditional cure, rubbing the frostbitten part with snow, can damage and even freeze more tissue.) External heat must be used very carefully: there is no feeling frostbitten tissue, making it very easy to burn. Burned frost-bitten tissue rarely recovers. The best way to warm a frozen extremity is to stick the frozen fingers or toes in an armpit.

Hypothermia is defined as having a core body temperature of 95°F or less. The body's temperature drops below the minimum needed to maintain the body's metabolic functions. Children and the elderly are more prone to hypothermia because they don't make heat as well as they lose it.

The early signs of hypothermia are shivering, loss of motor control in the hands, and shallow breathing. As hypothermia deepens, shivering increases, muscle coordination worsens, and confusion sets in. In the final stage of hypothermia, shivering stops but the victim is unable to perform even simple tasks and may become irrational.

The cure for hypothermia is to get the victim warmed up again. Even when some body functions have shut down, the preservative effects of the cold can sometimes make it possible to revive the victims with minimal effects. (There's an old rule in emergency medicine that no one's dead until they're *warm* and dead.)

Post-Blizzard Issues

When the blizzard's over, you're not safe yet, particularly if there's been a substantial amount of snow and ice.

Digging Out

It's not uncommon for there to be enough snow to bury a house, both from falling snow and drifts. Many houses in the Upper Midwest have access doors in the upper stories so you can get out of your house.

If you're digging someone out from the outside, dig from the top and work down. Eliminate the overhead risk and work your way down. You may even need to clear a portion of the roof.

Tunnels may look attractive, but they can collapse easily. There's no structural integrity to snow. A trough, similarly, will collapse on top of you with snow from overhead. Both are silly ways to bury yourself in snow.

If you're digging out from the inside, again, you'll be working with a small space outside. Suppose you open a window to get out: you're going to have to bring snow in with you at first. Always remember that you have to go across and up so that you don't create a tunnel that you go into and get crushed.

One of the hazards when snow drifts is a snow cavern. A snow cavern happens when the snow covers something without filling it. This could be something relatively small, such as a small drainage ditch, or it could even be large, such as a large hollow created in front of a house's front door. The danger of snow caverns is that they're unstable and won't support weight. You can fall several feet or more onto whatever is underneath.

If you're walking on a large accumulation of fresh snow, always carry a walking stick. Use the walking stick to poke through the snow and feel for a hollow underneath the surface. Some walking sticks will have a flat head on them so you can check the snow and see how far it'll go down, but the most important part is to find out where the bottom of the snow cavern is.

Flooding When It Thaws

Snow is nothing but frozen water, so when the weather warms, you may have a lot more water than is the norm. An exceptionally large snow-fall over the course of a winter may mean spring flooding. If you're in danger of excessive spring runoff, there will be plenty of warning on the news about the size of the snowpack and what to expect. Look at Chapter 7 for information on preparing for flooding.

Ice Damage

Ice causes damage because it's heavy and it builds up on things that aren't normally used to taking weight. Water expands as it freezes. As a coat of ice thickens, the ice expands in all directions, including inward. This, combined with the sheer weight of a heavy coat of ice, can crush things.

If something is frozen under a thick coat of ice, the amount of heat necessary to melt it free may make it impractical to remove the ice because it'll refreeze quickly as soon as you remove the heat. If something's that icy, you may need to wait until the weather warms up enough for the ice to melt on its own.

The Least You Need to Know

- The key to surviving any cold-weather disaster is insulation—for your home, for your business, and for your body.
- Heating your shelter requires forethought.
- A snow shovel is actually a manual snowplow. It is not designed to be used to lift snow, but to push it out of the way.
- Freezing temperatures and blizzards pose risk not only due to disruption of municipal services, but directly in the form of frozen or burst pipes, structural collapse under the weight of snow and ice, and even frostbite.

Chapter **13**

Pandemics

In This Chapter

- ◆ The difference between epidemics and pandemics
- ◆ Understanding the causes and effects of the most likely pandemic virus
- ◆ Immunization, prevention, and drugs
- ◆ How to know when the pandemic's starting
- ◆ What to expect in your community during a pandemic
- ◆ What life will be like after a pandemic

Of all the disasters discussed in this book, pandemic is probably the worst. Pandemics are the most far-reaching of all the disasters, having global effects and inconceivable death tolls. And yet, you can do a number of things to prepare for the coming pandemic that will minimize the effects and may even help you and your family stay alive and healthy.

In this chapter, we identify what a pandemic is and how it differs from an epidemic. We focus on influenza (the most likely and most predictable of all pandemic diseases), how the virus works, what you can do to avoid and prevent flu and other diseases, the effects that a pandemic will have, and how to deal with the social aftermath.

What's a Pandemic?

Thirty years ago, a pandemic was a disease that struck the whole world virtually simultaneously. (The Black Plague was a pandemic in the old definition.) An epidemic was something that might hit a region or even a continent, but it didn't hit the whole planet simultaneously.

Today, with the jet age, an epidemic is still seen as something that stays regional, so you could have an epidemic in North America that doesn't go anywhere else. But a pandemic can start here, then resolve or not, then move to Europe, and resolve or not, and so on. A pandemic is now a disease that can go anywhere but it doesn't have to be in all these places at once.

Several diseases could actually count as pandemics now, based on their global coverage: malaria, HIV/AIDS, and tuberculosis. These are all serious and fatal if left untreated, but they're not particularly immediate. Victims can and do live for quite a while even with no treatment whatsoever. A pandemic must not only be serious and potentially fatal, if must also be immediate.

The disease that is the single biggest immediate danger right now is influenza. Annually, there are between 35 and 50 million cases of flu in the United States. About 10 to 20 percent of the U.S. population gets the flu during the course of the year. Flu causes about 20,000 deaths directly and another 40,000 deaths indirectly from pneumonia and complications.

Some of the flu epidemics of the past would be called pandemics now, because they eventually covered the entire planet even though it wasn't simultaneous. There was one in the 1950s, another in the 1960s, and the swine flu in the 1970s. All these covered the whole world, but their genetic origin was different, so there was always partial immunity to the new epidemic strain.

The Day the Earth Coughed

Two major types of flu virus affect people. The first, type A, has a number of different subtypes, which are identified by two proteins on the viral sheath: hemagglutinin and neuraminidase, H and N respectively. These proteins are the keys by which the virus attaches itself to the body's cells. There are 16 different H types and 9 different N types. Type A viruses are identified by these two components, such as H3N2. The second type of flu virus, type B, is not divided into subtypes and is far less dangerous than type A.

Your body recognizes a strain of flu virus by the proteins on the viral sheath. If you've already had an H3N2 virus before, chances are that you'll have the antibodies to fight it off and at worst you'll only have a mild case. But if the flu virus mutates and becomes H3N3 or H2N7, your immune system can't identify it as easily and is less able to fight it off.

Not If, but When

Type A flu virus can go through two types of change. The first change is antigenic drift, where there are minor changes in the H or N proteins. These changes make the virus more effective at latching on to cells and reducing the immune system's response. Eventually, one of the proteins fits perfectly and the virus becomes very good at infecting people, causing an epidemic. This version sweeps through the population, who then develop antibodies for this particular variety of virus, and the cycle starts over. Antigenic drift is a largely evolutionary reaction in the virus that happens every seven years, plus or minus one.

The second change is antigenic shift, where new H and N combinations are created. For example, when two different flu viruses, such as H3N4 and H5N3, both infect someone at the same time, they can combine to produce a new combination, H5N4. Your immune system won't recognize this new virus at all. Because there's no immunity to a novel flu virus, it spreads very quickly. In addition, the mortality rate is much higher than it would be for other versions of the flu. The Hong Kong flu of 1968 and 1969 and the Asian flu of 1957 and 1958 were both antigenic shift versions of the flu. Antigenic shift happens roughly every 13 years, plus or minus 3.

The real danger happens when an antigenic drift combines with an antigenic shift. You then have a highly infectious virus (because the proteins attach to the cells very well) that everyone is susceptible to (because it's a novel virus). This, combined with the fast incubation period, lets a deadly version of the virus jump from person to person before they're immobilized or dead.

How often does this happen? You get a completely new type of flu virus when you get an antigenic drift and an antigenic shift at the same time. So every 91 (7 × 13) years, plus or minus 3 years, there's going to be a pandemic version of the virus appearing. (Other minor factors determine how nasty a virus is, but generally speaking, it's going to be bad.)

The last pandemic flu was the Spanish flu, which started in 1917. This means that the next pandemic version of the flu is due to make its first appearance in 2008, give or take three years.

As this book was going to press, there was an outbreak of H1N1 virus. This is a developing situation, and there will be information that comes available after this book has printed. For more information on what's currently happening with the H1N1 virus, see www.disaster-blog.com and www.pandemicflu.gov.

Not Who, But How Bad

When your immune system starts fighting off the infection, your cells start releasing chemicals (cytokines) that tell other cells to join the fight. Unfortunately, cytokines in quantity also cause the fever, muscle aches, and respiratory congestion that flu is notorious for.

For normal flu infections, children, seniors, and people with compromised respiratory or immune systems are at most risk of complications or death. Ironically, for epidemic and pandemic versions of the flu, the people at most risk are 20- to 45-year-olds in good health. The reason is that the normal at-risk groups have immune systems that aren't particularly competent at dealing with infection. However, the 20- to 45-year-olds have really vigorous immune systems that can fight bugs off really well—but they overcompensate.

If you think of your body as a battlefield between the virus and the immune system, having too vigorous an immune response can result in a "scorched earth" policy: a cytokine storm. Too much damage caused at the scene of the battle can interfere with normal organ function and cause organ failure and death. Most frequently, this damage affects the lungs, resulting in pneumonia and other fatal complications. In the Spanish flu pandemic, patients could go from healthy to dead of pneumonia in as little as a day.

We now have ventilators that keep people alive who would otherwise die, but these are specialized pieces of medical equipment for use by trained personnel only. If the numbers from the Spanish flu pandemic are any basis, here's what is likely to happen in the next flu pandemic:

♦ There are currently about 38,000 hospital beds available at any time during the flu season.

♦ Of the 105,000 ventilators in service, about 16,000 are available at any one time.

Based on the information available, about 33 percent of the U.S. population of 300 million is likely to get sick. Of those 100 million people, half will need assisted care of some kind. Half of those will need to be hospitalized and half of those will need a ventilator (and half of those will die). That's 12.5 million beds and ventilators.

The vaguely good news is that the 12.5 million isn't all at once. Instead it's going to be spread across 18 to 24 months. The bad news is that this is still an average of half to three quarters of a million people a month for 38,000 hospital beds and 16,000 ventilators.

Only One Risk Factor: Living on Planet Earth

Population growth is marked by the limits of human technology. Thanks to our larger brains and our opposable thumbs, humans have beaten the first limiting factor of population growth: predators and the natural environment. Humans spread across the planet and adapt to every climate available to them.

For the longest time, the thing that marked the limits of human technology was famine. The population would grow until there was a failure in our ability to produce enough food and then there'd be a famine. A lot of people who weren't famine-resistant would die. The ones who were left were better able to survive because they had metabolisms that either worked slower all the time or they would go into a slower metabolism when there wasn't a lot of food. The famine survivors would recover from the famine conditions, make changes to their food-producing technologies, and the population would start growing again.

Somewhere in the 1600 to 1800s, human food production got to the point that we could feed pretty much any size group of people. Famine was still a problem in places, but it wasn't the limiting factor to population growth it had been. And humans encountered the next failure point of technology: disease.

Hygiene and Sanitation Are Key

Disease has always been a problem wherever there are large concentrations of people. Thucydides wrote of plague during the Peloponnesian War, apparently caused by overcrowding in the city of Athens. Other plagues hit the Roman Empire and the known world in the second and third centuries before the bubonic plague outbreak in the sixth century. The bubonic plague continued to be a problem for the next millennium.

Most diseases have been a problem of hygiene and sanitation. (There's some overlap with famine, too. If you're sick, you can't bring in the harvests. If you're weak from hunger, you don't recover from diseases.) There was a growing understanding of how disease worked as time went on. Though there weren't cures for diseases such as TB and leprosy, humans figured out that these people needed to be *quarantined.*

def•i•ni•tion

> **Quarantine** comes from the Italian word for "forty." In Venice, they'd hold the ships for 40 days at cannon point to make sure that nobody died from plague. If the people on your ship survived, great. If not, too bad: the Venetians would take your ship out and sink it.

As antibiotics were discovered, diseases started losing ground and the population could grow still larger. This gave us the next big limiting factor: pandemic. The challenge with pandemics is to learn to control, prevent, and cure them. Flu vaccines and so on are solving this problem, but there's a long way to go. By extension, we're also seeing new and emerging diseases in the human population, such as HIV, West Nile virus, and SARS.

The Danger Is in the Dose

The key to avoiding the disease in the first place is to decrease your exposure. Even if you're not vaccinated, it's possible for you to develop immunity. The ideal of immunization is to expose yourself to enough of the bug to trigger your immune system.

By its very nature, a pandemic means you have a 100 percent chance of being exposed. But does it take a person coughing in a room once, or do you have to be there for a while? Suppose you have to be exposed to 20,000 viral particles in one sneeze to guarantee that you're infected, too. Then suppose that a blast of 1,000 viral particles won't infect you but will be enough to trigger your immune system to develop antibodies. You won't know what these numbers are for you and you certainly won't know how many viral particles you're being exposed to at any one time, but there's still something you can do. If you take steps to reduce your overall exposure, you stand a better chance of not hitting your infectious level while exposing yourself to just enough virus.

The key is to do two things. Most importantly, you need to decrease your personal exposure level by doing things such as bumping elbows rather than shaking hands, not sharing glasses, and washing your hands frequently. This reduces the likelihood of encountering viruses.

In addition, if you and everyone else cover their mouths when sneezing, the chance of exposure to stray viral particles is reduced. In fact, this can work out for a whole group: by reducing overall exposure, lots of people have a chance to build up immunity through exposure to just enough virus. (Along the same lines, if everyone in a group gets immunized, the number of people who can get the flu and then pass it on to other people is markedly reduced. This is known as herd immunity.)

A flu pandemic isn't 18 to 24 months of straight sickness. There will be waves of illness happening every six to nine months. If you develop a measure of immunity the first time it passes through, you may be able to resist it on the next passes.

An Ounce of Prevention Is Worth a Pound of Tamiflu

It's far better not to get a disease than to treat it when you've got it. Even with something as nasty as the flu, you can do a number of things to avoid it altogether.

Flu Shots

Flu shots are a good idea. The standard flu vaccine contains an inactive (killed) virus that triggers your immune system to develop antibodies. The types of viruses that appear in the vaccine are two type A viruses and a type B virus. (The type C virus isn't included in the vaccine because it's rarely life-threatening.) These viruses are chosen six to nine months before the vaccine is released based on information about the strains of flu that are likely to infect people.

You should get vaccinated as soon as the vaccine becomes available. It takes two weeks after vaccination for your body's immune system to kick into gear. Getting a flu shot early on will give you protection during the height of the flu season. (Children younger than nine years old who haven't been vaccinated before are likely to need a second dose in a month to make sure their immune systems figure out what to do.)

After you're vaccinated, you're ready for the flu viruses you're most likely to encounter for the whole flu season. You need to get vaccinated each year against the latest varieties of viruses. And even if you've been vaccinated, you can still get the flu in several cases:

- If you catch the flu while your body's still developing antibodies after a vaccination
- If the flu virus you're exposed to isn't the same one as you were vaccinated for
- If your immune system is compromised because of youth, age, or another condition (such as diabetes)

The good news is that if you get the flu after you've been vaccinated, the chances are good that it will be much milder than it would have been otherwise. This may not be much comfort when you're feeling sick, but it's true.

There currently aren't vaccines for the likeliest pandemic strains, H1N1, H1N7, and H5N1. There are more than a dozen companies working on developing vaccines at this time. Unfortunately, creating a vaccine for a new flu virus takes time. It takes at least two months for a trial vaccine to be created, then another four to six months for quantities of vaccine to be developed, another couple weeks to distribute them, and at least two more weeks for your immune system to respond. If a pandemic variety of flu virus suddenly appears, it would take a bare minimum of seven months for a vaccine to be developed and of any use. Meanwhile, people will be getting sick and dying.

Hygiene Habits

Flu viruses are transmitted either by respiratory droplets from coughing or sneezing or by touching something with the virus on it. When the virus enters your system, it multiplies rapidly and you start developing symptoms within one to three days.

The basic rules of hygiene and cold prevention you learned as a child are all the sanitation knowledge you need to keep from getting the flu. First, wash your hands. You should wash your hands whenever you do the following:

- Touch something that may have been infected.
- Go to the bathroom.
- Before and after dressing a cut, wound, or other break in the skin.
- Blow your nose, cough, sneeze, or brush your teeth.
- Touch something that's not itself reasonably clean (including garbage, animals, and animal waste).
- Before preparing food and after touching raw meat or fish.
- Before eating.
- After changing a diaper or cleaning up a child who's gone to the bathroom.
- Before and after touching anyone sick.

Most people don't wash their hands very well. Start by wetting your hands with clean running water. Use warm or hot water if possible. Apply soap to both hands, then rub your hands together until you form a lather. Scrub all surfaces on your hands, including under the nails, for at least 20 seconds. (Many people will sing to themselves: a couple verses of "Happy Birthday" or "Old MacDonald Had a Farm" work, but feel free to suit your own musical tastes.) Rinse your hands thoroughly in running water and then dry them completely using a paper towel or air dryer. If you use a paper towel, turn off the faucet using the paper towel.

Hand-sanitizing gels and creams use alcohol to sterilize. They're not as good as soap and water—your hands will be sterile but not necessarily clean—but they'll do in a pinch. Put a small squirt of sanitizer on your palms, then rub your hands together briskly. Be sure that you get sanitizer over all the surfaces of your hands and fingers and keep rubbing until the sanitizer dries.

Second, cover your mouth and nose when you cough or sneeze so you don't spread germs or viruses. For example, the flu virus is particularly hardy. It spreads primarily by being inhaled and the virus remains infectious for up to 24 hours after someone sneezes. Minimizing the droplets of virus people might encounter is a very good technique for preventing disease. Particles can travel as fast as 100 m.p.h. when you sneeze.

Covering your mouth and nose with your hands when you cough or sneeze isn't effective. You just create a cloud of droplets around your head. The best way to cover your mouth and nose is by tucking your face into the crook of your arm. This aims droplets down rather than outward.

def•i•ni•tion

A **fomite** is an inanimate object that can transmit germs or viruses from one person to another. Doorknobs, phones, faucets, toilets, banisters, steering wheels, keyboards, and desks are all fomites.

Finally, keep things clean. This includes anything you're likely to touch that someone else may touch (*fomites*). During flu season, you may want to carry sanitizing wipes, a pocket hand sanitizing gel, and a small container of Lysol or disinfectant spray.

Working at home is a good way to keep yourself away from potentially infectious people and things. If you're isolated from other people, you minimize opportunities to get sneezed on or near and to touch anything that was handled by someone infectious. Parents with children will have a harder time isolating themselves because kids that go to school will get exposed to everything fairly quickly and then bring it home.

All this is a lot of work, but it's actually fairly effective at preventing not only the flu, but the common cold, staph, strep, E. coli, and many other bugs that you're better off without as well.

The Real Deal on Drug Therapy

If you've got the flu, a couple of antiviral drugs do a fair job of helping the body destroy the flu virus. There are several problems, though. (You knew I was going to say that!)

First, there's not enough of these drugs to go around. Second, the flu viruses most likely to cause a pandemic are particularly nasty and will require double doses of both antiviral drugs simultaneously for twice as long as would normally be necessary. In other words, one case of flu will require four times the normal dose of each of two different drugs already in short supply to be effective. In addition, you have to start treatment promptly or it won't make a difference. Worst of all, the death rate is still going to be around 30 percent even with all this treatment.

Warnings

Chances are good that a flu pandemic will start by looking like any other outbreak of a serious flu. However, there will be more warnings at first.

News Reports

There will be news reports of major outbreaks of flu worldwide. It'll be similar to coverage of the SARS epidemic: there will be a lot of footage of really sick people, discussions of what is known about the disease, and reminders to wash your hands a lot and cover your mouth when you cough.

However, by the time we hear this news, jets will have carried it all over the planet. (Jets are a great way for flu to spread itself: the durability of the virus and the recycled air aboard planes make it almost impossible to avoid if you're in a plane with someone with the flu.)

Public Health Announcements

When it's clear that this is no ordinary flu, the next stage will be public health announcements. Mass gathering points and activities where lots of people get together will be discouraged or even closed altogether.

There will likely be reverse 911 calls, where the 911 system is used to do a series of robocalls to the phones in the area to provide health bulletins. The emergency broadcasting system will also be brought into play. Information will also be sent by mail.

Related Hazards

The related hazards of all the other disasters discussed in this book are about damage from floods, fires, wind, and so on. The hazards from a pandemic are all about people and the impact of absenteeism.

Loss of Public Services

Sooner or later, our technology has to interface with people. You have to have 911 operators, you have to have actual people provide emergency medical services, police, fire, and so on. Someone has to keep the power grid, the water, sewer, gas, oil, coal, and food flowing. The phones have to work and garbage has to get picked up. All these things require a human being somewhere. When absenteeism is running at 20 percent or better, there are going to be degraded services.

Depending on how bad things get during the pandemic, there may be mandatory shutdowns of services simply because no one is available to do them. Count on the cable guy never showing up. The phones may not work reliably. Stores of all kinds may not be open and when they are, they won't be stocked well. It's possible there will be food shortages because there aren't enough people to harvest the food, pack it, ship it, warehouse it, and stock it.

Medical Care Shortages

Health-care workers will show how the pandemic is going. They're at the most risk for exposure to the disease and their infection rates will be higher than the general population. In addition, because of the tidal wave of sick people requiring care and attention, you can also count on health-care workers working long hours until they're at the breaking point.

After the SARS epidemic in Canada, there was a very high resignation rate in the four hospitals that had dealt with SARS patients. Health-care workers left the field entirely because they were burned out or they had seen too many people die. This caused a serious shortage of medical personnel at all levels for a while. The same thing on a national or global scale could mean a shortage of medical care across the country and even the world.

Mortuary Care Shortages

Based on the statistics from the Spanish flu, the best estimates of mortality in the United States from a flu pandemic are roughly 2 percent of the population, about 6.75 million people, during 12 to 18 months. This is 10 times as many people as were killed in the U.S. Civil War. It's impossible to conceive of that many dead people, and it's not fun to think about. This many bodies will present serious problems. Apart from the emotional devastation, corpses are dangerous. You can pick up the flu directly from handling a body. In addition, bodies decay rapidly, causing a number of serious bacterial risks within a matter of hours. Reports from the Spanish flu pandemic talked about bodies stacked like cordwood on the streets.

There aren't enough graveyards to handle 6.75 million extra bodies in a year. The United States currently does about 850,000 cremations each year. Even assuming running three shifts, it's likely that as many as 4 or 5 million bodies would have to be disposed of safely. The only practical method of dealing with this many bodies is going to be mass graves.

Contingency plans have already been drafted that deal with this unpleasant possibility. The current plans discuss the use of ice-skating rinks or, in winter months in cold climates, outdoor cold to keep bodies from decaying until they can be transported to mass grave sites.

One body has to be buried at least 6 feet underground to prevent decay products from coming to the surface or accidental exposure of the remains from flooding or subsidence. But many bodies together require a much deeper grave to prevent decay and bacteria from contaminating the area. It's very possible that old mines may be pressed into service as mass graves, with a large concrete cap poured on top to seal the mine permanently. This will not be easy for anybody.

Social Breakdown

In a society where we expect a certain amount of service for our dollar, there's going to be a lot of frustration. During a pandemic, you can expect a breakdown of services and a certain amount of social unrest. Some of the direct effects will be the following:

◆ Businesses shutting their doors, initially on a temporary basis because of a lack of staff to run them, but then permanently because of a lack of customers or inventory.

◆ Slow response for public services of all kinds.

◆ Schools shutting down because it won't be prudent to expose children to the flu.

◆ Drastically reduced air travel because of the lack of people willing to risk being exposed to other people on a plane, as well as no airline staff who can fly the planes.

◆ Increased car travel on long trips, where you're only in your car and not exposed to other people.

◆ Periodic failures of utilities, Internet access, cable TV, and phone.

All these problems will require people to cooperate more, be more patient, and be a little more thoughtful of each other. People aren't good at this as a rule, but it's something to try. The old motto of "Courtesy is contagious" may be very helpful.

Post-Pandemic Issues

Even when the disease has burned itself out completely, there are going to be a lot of recovery issues.

Reconstitution of the Family

The normal social hierarchy is going to turn upside down. Because of the higher mortality rates for 20- to 45-year-olds, it's very likely that children will end up taking care of parents or performing parental roles, particularly if one or both parents dies. There have already been some examples of this: children in London during the Blitz tended to take over as parents and assume control of their siblings. In many cases, they were allowed to make legal decisions and they even directed older relatives when they were replacing patriarchs or matriarchs.

As things return to normal, one of two things will happen. First, the family may reconstitute with a bit of shuffling of the family roles. More distant relatives may take a role in the family. Everyone may move up in the family hierarchy, but they'll maintain their relative positions. However, the family may also form around the strongest wage earner as leader. What choices the family makes will largely depend on how strong the family is compared to their surrounding social structure.

Something that will affect many people is post-traumatic stress disorder (PTSD). PTSD isn't a problem reserved for veterans; it can affect anyone who's been through a terrifying experience that involves extreme physical or mental stress. People suffering from PTSD frequently feel numb emotionally, as well as having intrusive, frightening memories. Stress-related disorders will be very common, and it's likely that there will be community- or government-sponsored treatment facilities.

Reconstitution of the Community

During the pandemic, it's very likely that many political leaders, church leaders, and authorities at all levels will die, causing a breakdown in the community structure. For example, what happens if the president, vice president, and one third of Congress all die? Or, on a local level, what if the mayor and most of the city council die as a result of the flu? The same rules that apply for cooperating with other people, being gentle, and remembering that nobody will have a corner on grief and horror will apply for cleaning up after a pandemic.

Businesses will be particularly susceptible to the loss of people and to the history and knowledge they have. Fred may be the only person who knows which companies can manufacture a particular subassembly, how to start a production line, or how to maintain a critical piece of software. If Fred dies, the company loses that knowledge. If Fred's team of three engineers, who worked for the company for 15 years, also die, there's probably no chance of recovering this knowledge. Businesses that haven't institutionalized their knowledge and cultural history may suffer heavily or even go out of business as a result.

Most predictions suggest that the national and even the global economy will be in shambles, simply because production and distribution will have been crippled. This will take more work to get everything back together.

The Least You Need to Know

- ◆ Pandemic is a certainty. It is not if, but when and how bad.

- ◆ In a pandemic, you have a 100 percent chance of being exposed. The only questions are whether you will get sick and who will care for you.

- ◆ Washing your hands and covering your mouth properly are the best protection against pandemic flu.

- ◆ There aren't enough antiviral drugs for everyone and the doses typically given are not strong enough by a factor of four.

- ◆ When it comes to pandemic, the only approach is to plan ahead and prepare for weeks of self-sufficiency.

Chapter 14

Toxic Spills

In This Chapter

- ◆ Toxic spills: solid, liquid, or gas
- ◆ Where toxic spills are likely to occur
- ◆ How to evacuate and avoid exposure
- ◆ What to watch for if you've been exposed to toxins

A toxic spill doesn't just involve hazardous wastes, such as old nuclear fuel rods or 50,000 gallons of dioxin-laden liquid. A toxic spill also can involve something that is perfectly useful but is nevertheless poisonous or damaging to the surrounding environment. The Exxon *Valdez* released 11 million gallons of an otherwise highly marketable commodity into Prince William Sound, where it destroyed hundreds of square miles of beaches and ocean floor.

In this chapter, we discuss the different kinds of toxic spills, where they can happen, where to get warnings and information, evacuation, and the symptoms of serious exposure to several classes of toxins. We also discuss returning home and dealing with the possibilities of decontamination.

What Is a Toxic Spill?

Anything toxic—solid, liquid, or gas—can escape the thing it's contained in and create a spill. When it's a gas (or an aerosol version of a liquid or a solid), it's generally called a *release*. The toxic substance forms a cloud, but it's still a toxic spill.

When a solid spills, the distribution depends on its weight, the forces acting on it (such as water, wind, and gravity), and the particle size. A light, dry chemical can ride the wind as a dust, whereas large chunks—bricks of chemicals—can fall and just sit there. If the solid dissolves in water or evaporates, it can spread; a shipment of solid pesticides may get picked up by a stiff breeze and blown a fair distance from the site of the spill, then dissolve in water and spread even further. A spill might even be a truckload of mothballs, which can scatter like large gravel, but if they land on hot asphalt in summer, they're going to turn into a heavier-than-air poisonous cloud.

Liquids can mix with water or not. If they're miscible, they'll flow with water. They may aerosolize for some reason—for example, alcohols mix with water but will evaporate fairly quickly—but until then, they act like water. Nonmiscible liquids are things such as oils and other petroleum products and many solvents. If they're lighter than water (as most refined oils or gasoline), they'll float on the surface of water and may evaporate or burn before eventually breaking down. Liquids that are heavier than water will sink. When crude oil hits seawater, most of it sinks below the surface of the ocean and rests on the ocean floor. (Some parts of it separate and float on the surface for a while.)

If a release of gas is lighter than air, it'll disperse with the wind and float off and up, probably out of harm's way. Natural gas is slightly lighter than air, so it will rise (slowly). If the gas is heavier than air, though, it acts like a liquid. It will stay close to the ground, but it won't sink into the ground like a liquid does. The gas will, however, fill ditches, depressions, and contours, and flow like water.

Spills on Water

The characteristics of the chemical that spills determines how the chemical is going to act in the environment it spills in. A toxin might act differently spilled on hot, dry ground than on cold, wet ground.

A spill in water can add many complications to the spill. For example, crude oil both floats and sinks. Things that dissolve in water will generally do so, but there may also be too much chemical for so little water that it all just forms a sludge of damp toxins. Powders may dissolve in water or not. The insoluble ones are like sugar in gasoline, where it sinks to the bottom. The soluble ones can dissolve slowly or quickly, similar to sugar in cold or hot water.

Gasses are either soluble or insoluble. A soluble gas, such as raw chlorine, that's released underwater may partially or completely dissolve in the surrounding water as it bubbles up. However, insoluble gasses, such as carbon monoxide, will bubble up through the water to the top, where they'll act as a lighter- or heavier-than-air toxic spill.

Spills on Land

Spills of solids or liquids on land act about as you'd expect: they spread by wind, gravity, or other liquids.

Gasses that are heavier than air follow the contour of the land to the lowest points (just as liquid does). The danger is that they can pool. By the time you've noticed them, it may be too late. For example, slow fermentation in silos can create a buildup of methane or carbon dioxide, both of which are heavier than air. The bottom 10 or 20 feet of air in the silo can be displaced. Raw methane is toxic, but walking into the silo can be fatal simply from the lack of oxygen. Lighter-than-air gasses disperse quickly in a larger volume of air, so the toxicity of the spill depends on the concentration of the gas.

Liquids on land flow like a heavier gas, but they can also penetrate the soil. This adds the very real danger of contaminated groundwater, making it a water spill that can be transferred by underground rivers and aquifers.

Spills Underground

Many underground spills are pipeline ruptures of one kind or another. In these types of spills, the gasses or liquids are under pressure. This means that an underground spill is not merely a spill of material, but the chemicals are also being injected into the surrounding soil under pressure.

Another kind of underground spill is when a solid, liquid, or gas is deliberately disposed of. This can be illegal dumping of a bunch of pesticides in the local landfill, or a corporate disposition of a lot of toxic wastes.

Everytown, USA

Toxic spills can happen virtually everywhere. Between commercial trucking, trains, warehouses, and pipelines, there is no municipality that doesn't have hazardous material (hazmat) transported through their region in substantial volumes. The question, therefore, is not if you're at risk or not, but what is being transported or used.

Toxic spills can also be related to industrial waste. Many industries use incredibly toxic substances as part of their everyday operations. If the company has not been disposing of their wastes properly or simply goes out of business, there can be a hazardous accumulation of chemicals. This can be a very large accumulation, such as an EPA Superfund cleanup site, but small waste dumps can present real problems to communities as well.

A plating company in Seattle had been in business for decades. Plating companies use large quantities of cyanide as a key ingredient of the plating process. The company had been a reliable corporate citizen and had disposed of its waste cyanides in accordance with the law. In the 1980s, new stringent regulations on waste disposal went into effect and the company chose to close its doors. However, there were still large quantities of cyanide polluting the ground all around the business's old location—and nobody to clean it up. There are many hazardous waste sites similar to this, both known and unknown.

Many companies can't have their hazmat removed because it's expensive or the local cartage company that used to do it is out of business. In some cases, manufacturing companies adopt a method that allows them to have intermediate hazardous wastes. They bring in one hazmat that they turn into something that's different and a smaller quantity, then turn this into yet another, smaller quantity of a third type of hazmat.

For example, you'd start with 1,000 gallons of hazmat A, which you then use to create 700 gallons of hazmat B. From that, you create 300 gallons of hazmat C. It may be most cost-effective to ship hazmat C when you've got a large quantity of it, so you'd store the 300 gallons until you've got enough to make it worthwhile to ship. Meanwhile, the hazmat is sitting there on your plant site waiting to be shipped. Even when a company is acting responsibly, storing its hazardous waste and disposing of it appropriately, there's still a quantity at the plant most of the time.

What's in Your Community?

The good news is that what hazardous materials are in your community is public information. Users and transporters of hazmats of all kinds have to file information about what they're doing and when.

You can also determine the types of hazmat being used by manufacturers and businesses in your area. Permits are required for use and storage of hazmat in virtually any business. Depending on what kinds of hazmat it is, there may even have to be a set of public hearings for the initial permit.

What's Passing Through Your Community?

The bad news is that, although you can get the information about where it's being picked up and when and where it's going to be dropped off, you don't know the route. Trucks usually go via the shortest route to minimize the mileage, but other factors can change the route.

Similarly, freight trains schedule trains based on the traffic on the lines. A delay in Colorado can trigger rerouting of trains as far away as Michigan. There's just no predicting what's going to be going where, when.

Information Is the Only Preparation

The only way you can truly prepare for a potential toxic spill is to know what's in and what's coming through your community. The problem with knowing what's coming through, though, is that there are too many things that are moving from A to B. Even if you had a complete database of everything that was moving and by what route, you'd still spend all your time preparing for accidents.

Prep Facts
You can get warnings about what's traveling through your community. The color, size, and shape of the placards on trucks and the sides of tanker cars tell you the type of chemicals inside. You can look this information up online. You can also buy a copy of the Emergency Response Guide (aka "the Orange Book") at any truck stop. (Truckers need to know what signs to put up on their trucks.)

All kinds of horrible things might happen when materials are being transported, but they probably won't. These are unpredictable and unforeseeable accidents: a supertanker going aground, a train derailing, or a tanker truck having an accident. These things generally occur as the result of some technological failure or human error. Anytime materials are transported, it's a separate opportunity for a spill or release, but the likelihood of an individual shipment having any kind of problem is statistically almost zero.

Even though it's not practical for you to know where everything is on the road at any time, you can know where the fixed hazards are. It's not too hard to identify the major manufacturing plants, warehouses, and storage facilities in the community that have hazardous chemicals and wastes. They're always there and there's a finite chance of an accident occurring at some point. And because of that, you can prepare for them. For example, if you're downwind from a water-purification plant that uses a lot of liquid chlorine gas, you should be ready for an announcement that there has been an incident.

Warnings

Toxic spills (particularly large toxic spills) are reported similar to radiation hazards: they're going to be on every possible communications channel.

Radio and TV

Local radio and TV will cover toxic spills heavily. You'll be warned about what kind of spill it is, what's being done to contain it, and how to avoid being exposed to it. In addition, if the spill is particularly dangerous (such as a trainload of pesticides or a leak of 1,000 gallons of bromine from a nearby plant), there will be information on the symptoms of toxicity and what to do in case of exposure.

As with many other hazards, a toxic spill may activate the reverse 911 system. There will be instructions on whether to evacuate or stay indoors, where to tune in for information, and what to expect from clean-up efforts.

 Storm Warnings

If you're in an area where toxic gasses could drift downwind quickly, evacuate immediately even if there's a stiff wind blowing the gasses away. Winds can and do shift suddenly and unpredictably.

Sirens and Alarms

If the toxic spill is because of an accident at a plant or storage facility, there may be an emergency siren or alarm. A warehouse fire may only mean sirens from the fire engines. Any siren or alarm you can hear inside the house probably means you're way too close to the spill.

Generally speaking, if you're upwind, you're fine—as long as the wind doesn't shift.

When the Alarm Sounds

You've been alerted to the fact that there's a toxic spill nearby. Before anything else, get everyone inside, close all windows and doors, and turn off your air-conditioning or heating equipment. (Don't forget to close chimney and furnace dampers as well.) This keeps any fumes or fine particulates from entering the house and causing contamination.

It Is Never Soon Enough ...

... to get the heck out of the exposed area. Before you evacuate, though, you need to know where the spill is and whether evacuating will expose you to the spill. If the spill is actually a release of toxic gas and that cloud is already outside, don't go outside. There is a certain amount of air in the house and your structure will provide a semi-permeable bubble of safety.

Know Where to Go

After you identify the likeliest source of toxic spills, you also need to identify how far away these sources are from you and if you tend to be upwind or downwind of them. Accidents can happen pretty much anywhere, anytime, but the basic safety procedures for toxic spills are the same as for radiation accidents:

◆ Stay inside until instructed otherwise.

◆ Turn off the air-conditioning and heat right away so you don't pump contaminated air inside.

◆ Shower off immediately.

◆ If you have to go outside, dress with hoods, long sleeves, pants, and gloves. Tape everything down so air can't get in.

Listen to the Pros

People who deal with hazmat have a grim sense of humor. For example, here's the "Copological Hazmat Identification Method" used by firefighters. If there's a toxic spill of some kind, look for the police officer standing by his car.

1. If the officer is standing and the car is running, the spill is not immediately hazardous.

2. If the officer is unconscious and the car is running, there are toxic fumes.

3. If the officer is unconscious and the car is stalled, there's an oxygen-displacing chemical in the area.

4. If the officer and the car are both melting, there's a corrosive chemical in the area.

5. If the officer and the car are on fire, the spill is extremely flammable.

6. If the officer and the car are glowing, the spill is radioactive.

However, if this is too complex, there is an even simpler method of assessing danger used by EMS crews. They look at the stripe on the

police officer's uniform slacks. If the stripe is vertical, the scene is probably safe. If the stripe is horizontal, it's not.

Related Hazards

All toxic spills are by definition poisonous, but what kind of hazards
any individual spill presents will vary depending on the substances and
circumstances.

Vapors, Clouds, Fogs, and Flows

Although it may not be as dangerous long term, anything that is a gas
or that vaporizes is going to present the most potential danger at first
simply because of the speed and invasiveness of gas clouds. For example,
a train wreck near Graniteville, South Carolina, in 2005 released 90
tons of chlorine gas from a single ruptured tank car. Nine people died
and approximately 250 were treated for chlorine exposure. The accident
required the evacuation of 5,400 people from the mile surrounding the
accident. Getting away from the area quickly was all that was required,
but speed was essential.

Things that come in clouds are likely to be simple chemicals, such as
chlorine, bromine, or other gasses. These are certainly dangerous and
caustic: chlorine and bromine, for example, have both been used in
warfare. The damage from such chemicals tends to be direct, physically attacking tissues that are exposed to it and causing toxicity from
overexposure.

However, the gas may also come from a fire that is sending up clouds
of more complex chemicals, such as pesticides or petroleum solvents,
which do not normally get distributed as gasses. These can expose you
to powerful and dangerous toxins that can cause far more damage.

Toxidromes: Signs of Trouble

When the body is exposed to a toxin, it exhibits a certain stereotypical
set of symptoms as the toxin infiltrates the system. This set of symptoms is known as a toxidrome.

Be on the lookout for two particular toxidromes: one for anticholinergic agents and the other for cholinergic agents. Anticholinergic agents suppress the release of acetylcholine in the brain, whereas cholinergic agents liberate it. Both of them have a direct effect on your body's nervous system. You don't really need to know this except to know where the terms come from.

Anticholinergic agents include nerve gasses, pesticides (but, interestingly, not herbicides), and certain plant toxins such as belladonna. Although usually immediate, the toxidrome for anticholinergic agents may take up to six hours to manifest. (The longer it takes to evolve, the more likely you are to die from the exposure.)

There's a mnemonic that doctors use to describe the toxidrome for anticholinergic agents: "Mad as a hatter, dry as a bone, red as a beet, hot as a hare, blind as a bat." These relate to the following symptoms:

◆ Mad as a hatter: You act irrationally or are delirious.

◆ Dry as a bone: You dry up and your mouth gets dry (like being on a load of antihistamines).

◆ Red as a beet: You're flushed and you stop sweating.

◆ Hot as a hare: Your heart rate's up and because you're not sweating, your temperature is up, too.

◆ Blind as a bat: Your pupils constrict to absolute pinpoints, because the effect of the chemical has paralyzed them.

Other subsequent symptoms of anticholinergic agents are your gastrointestinal tract, heart, and respiration slow.

Cholinergic agents, on the other hand, include compounds such as nicotine and muscarines (the poisonous alkaloids in mushrooms). The mnemonic for the toxidrome for cholinergic agents is "SLUDGE." This is an acronym, standing for the following symptoms:

◆ Salivation: Drooling

◆ Lacrimation: Tearing, runny eyes

◆ Urination: Increased urination from your bladder contracting

◆ Defecation: Diarrhea

- ◆ Gastro-intestinal upset: Extreme nausea
- ◆ Emesis: Vomiting

Anticholinergic and cholinergic agents are the two biggest dangers. They are much harder to treat than simpler compounds such as chlorine or hydrogen sulfide, because they're more invasive and they cause complex damage. If you're exhibiting either of these toxidromes, you've received a serious and possibly fatal exposure to a toxin. You're going to be in bad shape if you don't get antidotes. You should seek medical assistance immediately.

The real key to toxidromic chemicals is that they're attacking nerves. Pesticides, for example, are used for killing complex creatures such as cockroaches or rodents. (The only difference between VX nerve gas and pesticides is you're dealing with a bigger, badder creature.)

After the Spill

Even though the worst part of a spill may have been cleaned up, it's very possible that you still won't be allowed to return to your house. The amount of decontamination necessary will depend on the type and quantity of chemicals that spilled.

Home Decontamination

In many cases, the structural decontamination resulting from a toxic spill is cleaned off by rainfall. (The solution to pollution is dilution, after all.) It doesn't matter much what the agent is, as long as you dilute it, so water's almost always a good idea in large quantities. However, there are also circumstances where officials will come and decontaminate structures depending on the contaminant. This is not a job for you, the homeowner. It's incredibly complex and you need a technician to do this for you.

One possibility in the case of more dangerous decontamination is that the municipality will decontaminate. Another is that the contaminated houses might get leveled. It depends on the chemicals involved and the funding available.

Scrap and Bury—Reclaiming the Land

Some chemicals are easy to clean out. Some, such as chlorine, will cause significant damage to the environment, but they will also dissipate with relatively little lingering toxicity. Other chemicals, however, may be extremely difficult to remove. Some solvents, for example, can remain embedded in an area's water table for years, causing cancer, birth defects, and nervous system problems. Depending on the spill and the level of contamination, the only solution may be to scrap what's there and bury everything under a thick layer of clay.

Lingering Medical Effects

There may also be lingering medical effects to consider. This depends on how toxic the chemical is and how well the chemical bonds with the body. Some chemicals are easier for your body to bond with than others. Some chemicals can be flushed out of your system quickly with little damage, while others may also "age," where they end up bonding with your body permanently. If you undergo a chelation treatment quickly, you can clean some of these out, but maybe not.

The Least You Need to Know

- ◆ The solution to pollution is dilution.
- ◆ Decontamination is the best treatment for toxic exposure.
- ◆ How a toxic spill behaves depends on the characteristics of the material spilled.
- ◆ How a toxic spill moves depends on gravity, water flow, and wind direction. Upwind and uphill are the best places to be when near a spill.
- ◆ Classic antidotes and symptom patterns are actually rare in toxic spills.

3

Post Disaster

When the disaster's over, you need to return to normal life as quickly as possible. If you evacuated, you need to return home and do any cleanup necessary. You'll also need to get your business back to normal quickly. Damages to the house need to be dealt with through insurance or disaster aid. Even if you failed to plan, there are still things you can do that will give you a leg up and improve your chances of survival and a quick recovery.

When everything's over, it's time to figure out what worked and what didn't by evaluating your successes and failures with your disaster plan. See where you can improve and you'll be that much more prepared the next time.

Chapter 15

Getting Back to Work

In This Chapter

- ◆ Your employees are your most important asset
- ◆ The need for safeguarding and sustaining
- ◆ Partnering with the employee
- ◆ Restoring the community
- ◆ Conducting business ethically

The natural instinct of people during disaster is to first gather those most dear to them, then partner with those most useful to them. The window of opportunity for a business to be useful or even relevant is vanishingly small after a disaster strikes. However, if the business has built relationships with its employees and customers, it can make sure its employees and customers partner with it because it's useful to them.

In this chapter, we walk you through how to get your employees to come back to work even in the middle of a disaster, what the employees want from you, how to become a member of the employee "tribes," what you can do to support them, how to extend your support to the rest of the community, and how to provide volunteer services without losing money.

Business During a Disaster

During a disaster, the most important thing that a business can have is employees who show up. All things equal, the business that is able to mount the best response is the business with the most people responding. You, as the employer, need to have your employees come in and help you deal with any disaster-related cleanup and then run the business as best you can under difficult circumstances. The challenge for you, the employer, is to give your employees the same desire. They want to stay with their *tribes:* their families and the people close to them.

def•i•ni•tion _____

> A **tribe** is usually someone's family, but it can include any combination of relatives, neighbors, friends, and even pets. The tribe is the group that a person is emotionally tied to. Tribes tend to be small (usually less than a dozen) and are probably geographically close as well.

Even if the employee is returning to work, the employer has to allow for the fact that employees have another full-time job: organizing and managing the recovery of their own home and their tribe. This is the first conflict: your business needs manpower and your employees' tribes need them. The individual employees must choose where to place their attentions, and, not surprisingly, the business is positioned to lose this battle. The trick for businesses is to include the employees *and* their tribes, so that the employees don't have to choose their tribe over the business.

To get employees to show up at work means that the employer must change perceptions to the idea that employees can allocate manpower to the tribe and the business simultaneously. To do this, the business needs to partner with the employees by fulfilling the employee's survival functions and safeguarding and sustaining the employees' tribes.

When you're dealing with people who have been displaced by a disaster and have been stripped of their most basic habits and psychological needs, the tendency is for people to seek to re-attain some measure of security and normalcy, and then develop emotional inertia. This can be so great that you frequently have to almost blast people out of shelters. They've gotten comfortable where they are and feel settled, so even when offered hotel rooms or a FEMA trailer, they choose to stay in the communal shelters. It's not uncommon for shelter management to need to bring in law enforcement to encourage people to leave, not because they're violent but because they're just not moving. The uniforms are very effective at overcoming the emotional inertia.

Employers have a huge opportunity here. They should arrange for alternative housing as part of their disaster planning, even if it's only "we're providing stuff to the Red Cross and we'll all be volunteering." By doing this, the employer is providing an island of safety and is once again reinforcing the idea of the employer as a valued part of the employee's tribe. This in turn creates strong social bonds that last for a long time with the employees and their families.

Safeguarding and Sustaining

Safeguarding one's tribe is the most basic human function. Before an employee or customer will leave the tribe and go to work, they must be assured that they and the tribe are safe in a very concrete fashion. The businesses need to encourage the individuals that they're going to be safe and secure at the business site and that, by coming there, the individuals will increase the safety and security of the rest of their tribe, regardless of whether the tribe comes to the business site or not. This is why, even before roads are safe, people will venture out to purchase building supplies at the local hardware store or obtain medical care at the emergency room. Similarly, hospitals and emergency shelters have long known that if they allow the families of their staffs to shelter in place at the work site, absenteeism is dramatically reduced.

Here are just a few of the things that a business—any business—can do to help safeguard employees and their tribes:

- Shelter assistance
- Home-repair services
- Alternative housing sites
- Travel assistance
- Transportation
- Disaster daycare

Some companies hold "family preparedness fairs" at work, where each employee and their family learns to create a family disaster plan and to prepare themselves and their homes for disaster. These contribute to an ongoing sense of safety as well as underscoring the message that the employees and their families are important to the company.

The key for businesses becoming a partner with their employees' tribes is to become essential for sustaining the tribe's members. In times of crisis, tribes seek to become similar to spacecraft: self-contained vessels with air, water, food, clothing, and fuel. The business must attract individuals and customers by helping sustain the tribes.

This can be done in many ways. For example, few hardware stores are in the business of selling canned foods and camping meals, but when a hurricane threatens, these stores stock the shelves with water and canned goods. These will attract customers already in need of the safe-guarding that building repair supplies facilitate.

Most supplies that sustain the tribe are easily stockpiled and even more easily distributed. Remember, though, that these supplies only attract people if they believe their tribe is safe. Moreover, these sustaining supplies only attract people to a particular business if that business is associated with the tribe's safety as well. One possibility is that, rather than stockpiling food and water at the workplace, the business might have each employee stockpile their share at home. During a disaster, their Go-Pak can be replenished each evening after work from that stockpile and if the family is in need, the stockpile can be shared without risk of repercussions.

One of the best examples of safeguarding and sustaining came after Hurricane Charley, when a major Florida coastal hospital suffered roof damage. Other than a few wet walls and a leak in the stair well, the hospital was dry and secure. Despite this and a contractual requirement to report to work, a disturbing percentage of employees did not come to work for several days. The employees were busy safeguarding them-selves and their families. The lack of physical security intensified the need to stay connected. Family tribes stayed together repairing homes, buying generators, and even buying food as a group.

The hospital investigated the causes for the high absenteeism rather than just firing critical people summarily. The hospital learned that 94 percent of their employees lived within 1 mile of the coast and

99 percent of all employees had roof damage. In response, the hospital convinced their roofing contractor to give first priority (after the hospital) to hospital employees, beginning with those who worked during the disaster or the disaster response. This cost the hospital nothing and guaranteed the roofing contractor paying business while ensuring employees would have home repairs completed quickly. A true win-win-win situation.

The very next hurricane (Francis) caused more roof damage than Charley, but there was virtually no absenteeism following the storm. By providing for the safety of their employees and their tribes, the hospital's needs became essential to their employees.

Ensuring Emotional and Spiritual Comfort

It's not enough to safeguard and sustain people and their tribes. Businesses must also provide emotional and spiritual comfort. Studies of disaster survivors have shown that constant news feeds and updates about a disaster result in personal isolation, information overload, and severe disaster-induced stress. Before people can process information about a disaster effectively, they must feel reasonably calm and grounded. TVs and radios continuously tuned to disaster-related news serve as a constant and painful reminder of the disaster.

The emotions of employees, customers, and their tribes during and after a disaster determine whether they'll continue to return to a business. These people are often unaware of why they choose not to return; all they know is that they don't feel comfortable or safe. During and after a disaster, people want to get back to normal as quickly as possible. The business capable of providing a "normal" environment can be a huge source of comfort for employees and customers.

Storm Warnings

Empower the employees to help themselves by helping you. Make participation in the operations plan easy rather than mandatory. Also make reporting to work on "disaster day" an achievement unto itself and provide copious praise for the effort.

If the workplace is reasonably secure or has safe storage facilities, the business should encourage employees to keep personal photos and mementos in the workplace so that they feel a personal emotional connection to the workplace as a home away from home and their colleagues as their family away from home. Also make arrangements for rest areas and cots for employees who may have to stay on-site after a disaster and consider installing a generator or other means of maintaining a comfortable work environment. Having heat or air-conditioning at the workplace when these may be unavailable for the employees at their houses can be a powerful incentive.

Maintaining Interpersonal Connections

Disaster spurs people to seek connection to their tribe and others who have shared the experience. Being able to reach out to those you're closest to is critical. Keeping employees and customers connected to their tribes will make them feel more connected to their tribes and to the business. Even if the employees don't use it, the phone says, "You are important to this organization and so is your family."

Some of the things businesses can do to assist their employees and customers to maintain connections are to …

◆ Allow personal long-distance calls to people outside the area.

◆ Provide e-mail opportunities for employees and customers.

◆ Give employees pagers or prepaid cell phones for keeping in touch with family members.

Building for Tactical and Intellectual Success

It's not enough to just support the tribe's intangible needs. A business must also contribute to the tribe's ability to become self-sufficient again. The employee and the tribe must gain skills and knowledge to prepare for and respond to a disaster. In addition, as the employees and their tribes respond to a disaster, the business must promote and maintain a connection to the community.

The employee must be able to adopt these skills at home to ensure the uninterrupted operations of the tribe. Employers can best contribute to the employee's tribe by making sure that the disaster preparedness skills they learn at work can be used at home as well. Fortunately, the majority of business-continuity skills are directly applicable to the employee's home, family, and tribe, such as ...

- Securing critical data and original legal documents in an off-site location, such as a safe-deposit box.

- Regularly backing up computer records and even family photos onto CDs or DVDs.

- Performing yearly reviews of insurance coverage and physical home security.

- Reviewing and drilling the family disaster plan at least twice a year.

Restoring the Community

As communities and local social life reconstitute themselves in the aftermath of a disaster, most people will likely participate in the recovery. The bonds built during this period are often stronger than those that existed before the disaster. (An example of this is the close friendships that start in evacuation shelters.) Businesses that develop relationships with employees and their tribes will become a part of every societal bond that an employee or customer and their tribes make during the recovery, resulting in greater loyalty by employees and customers.

On a longer-term basis, businesses that promote the recovery of the local community and society never go unnoticed. In more than a quarter century of disaster response and recovery work, I have never encountered a business that went out of its way to support the community recovery that didn't benefit directly through new customers, returning customers, and long-lasting customer loyalty.

Volunteering vs. Business

When a disaster strikes, many businesses and individuals are eager to volunteer and assist those in need. They want to help rebuild the damaged homes and businesses, and they often donate the necessary materials and manpower to do so. Unfortunately, the resources that arrive on a volunteer and donation basis typically run out much sooner than expected. Very often, those businesses who gladly gave their time and resources to those in need feel guilty charging for additional services when their volunteer time ends. So they pack up and leave the area, proud of their good deeds, yet leaving those in the disaster area with few recovery options.

A great example of this is what happened in Port Charlotte, Florida, after Hurricane Charley. Initially after the hurricane, a large number of contractors went to the area, donating services, supplies, and other things needed to rebuild the community. The federal government also came in and paid many of the rebuild bills, even things not normally covered by FEMA. Then the money started to run out, and the majority of the volunteers went home. The residents of Port Charlotte did not want the contractors to leave and would have paid the contractors their normal rate to stay and finish the disaster recovery efforts. It was the contractors who originally came on a volunteer basis that felt guilty about taking money from disaster victims. As a result, even two years later, many Port Charlotte residents were still seeking reputable contractors to help them rebuild. It is an unfortunate situation that does not have to happen.

It is possible for businesses to maintain an income during a disaster and not feel guilty. In fact, those businesses that are able to help a community after a disaster strikes and offer a needed product or service can even profit handsomely and ethically.

Ethical Business Practices

When it comes to profiting from disaster, most people think of price-gouging or profiteering. Price-gougers seek to profit from misery, but price-gouging is not only immoral and unethical, it is also illegal. When Hurricane Katrina hit New Orleans, unscrupulous storeowners sold generators that normally retailed for a few hundred dollars for several thousand dollars. All those people received jail time or fines, and their business licenses were revoked.

There are three ethical ways to make money after a disaster. Volunteer your time and donate your products or services. You cover all your own costs and accept nothing in return, other than perhaps food and lodging. In return for your time and materials, you get the warm, fuzzy feeling of doing something good for the community. You become an everyday hero. If you are visible during this time, you also get great publicity, which could lead to business down the road from those who remember your good deeds.

Offer your products and/or services to the community at a discounted rate. Those who go this route figure out how low they can price something without putting a burden on the business. Realize, though, that no one in the community asked for the discount (although none will turn the discount down, either). Often, the business owner gives the discount because he or she has some level of altruism.

Come into the community and bid a fair market price for a product or service, roughly equivalent to what other companies would charge during nondisaster times. This is the key to ethical pricing. As long as you're not jacking up your prices simply because you know that people are desperate, you aren't price-gouging. And because it's a fair market price, people are more than happy to pay it. If your own supplies are more expensive—for example, if the cost for building materials is twice what you normally pay—you will need to pass this on to the customer, but tell the customer what the difference is and why, and don't charge a penny over your increase. This is completely moral and ethical. Unfortunately, few businesses make the transition to full price after starting out as a volunteer. However, if you really want to grow your business, this is the way to go.

Going from "Free" to "Fee"

The ideal situation for every business in a disaster-stricken community is to move from volunteerism back to regular business operations. However, most businesses aren't sure how to do this. Here are some suggestions for strategies on curtailing your volunteer activities and going back to being a paying business:

First, be upfront about how long you can offer your products or services for free. Tell people, "I can afford to volunteer for two weeks. I can afford to bring X amount of materials. If we run out of materials before two weeks, you supply the material and I'll stay the remainder of the time I stated." After the two weeks are up, talk with the people you have been helping. Explain again that you could only afford to volunteer for two weeks. Very often, at that point, they will ask you to bid the remainder of the work. Then you can offer a fair-market bid. If you get a "yes," then why would you not stay? You are already there, and now you are making money. If they say "no," then they are taking responsibility for their own recovery. At that point, you can go home and tend to your business, knowing that you have done a good deed.

Prior to a disaster, when your community is preparing its disaster-relief plans, put your company on the list of businesses available to aid in the recovery efforts. Businesses can work with their local communities to be "first-called" in the event that a disaster strikes. This means that, assuming your business is capable of responding, your business will put the community recovery at the front of the line.

Be proactive and contract to provide recovery services before a disaster strikes. A client company contracts with you and pays a retainer fee so that in the event of a disaster, you will put that company at the front of the list. In return, you get the contract for the other work the company needs done. This is completely ethical. In fact, it's a win-win solution. The company gets the repairs they need done and you have guaranteed work.

The Least You Need to Know

◆ Businesses need to safeguard and sustain their employees, customers, and their tribes, or they'll be alone in a disaster.

◆ The business that has planned for a disaster can build more and stronger connections with its employees and customers.

◆ After a disaster, people are happy to pay you a fair price for your products and services and it's okay for you to charge them.

◆ Businesses can help people and communities in many different ways. Volunteerism always pays off in long-term customer satisfaction and repeat business.

Chapter 16

Claiming Your Due—
Doing Your Best

In This Chapter

- ◆ Finding information about disaster aid
- ◆ Sources of assistance: local, state, and federal
- ◆ Getting your paperwork together for insurance
- ◆ Maximizing your insurance payout
- ◆ Winning with the insurance company

You've made it back home, looked everything over, and assessed the damages. Now you need to know how to take advantage of assistance from various governmental and nongovernmental agencies, as well as your insurance company.

In this chapter, we provide an overview of the maze of aid available from assistance organizations, insurance companies, and government agencies. In particular, we give you some insight into how insurance companies work in a disaster and how you can use this to your advantage.

Information Without Overload

Your government really is here to help you. There are agencies at all levels whose sole function is to provide assistance to people who are experiencing or recovering from disasters. The big trick is finding them so you can take advantage of what they have to offer.

There are currently more than 1,200 federal agencies of all kinds. In addition, there are state agencies to sort through, typically about 200 or so, but populous states such as California can have more than 500. On top of that are regional, county, municipal, and even neighborhood agencies that may be involved in the disaster aid and recovery process, as well as law enforcement agencies, utilities, and municipal services such as hospitals and firefighters. Add to this the many nongovernmental organizations (NGOs) that deal with disaster and recovery at all levels, such as the Red Cross, churches, and other groups from local to international, and you've got a lot of resources for finding the assistance you need to get back to your normal life.

Prior to an emergency, you should research what's available in your area for disaster relief. Many larger municipalities and states have an Office of Emergency Management or the like. These offices coordinate disaster relief at that level and can provide extensive information about what to expect for disaster relief.

Before you're hit by a disaster, the best place to start is the web. There are a number of websites devoted to providing information about who to talk to and where to go. Many of these are listed in Appendix A.

FEMA, Forms, and Funds

There are three kinds of governmental agencies: local, state, and federal, each of which has a different set of rules and a different focus. The one thing they have in common is that whenever a government agency is handing out money, it's a pretty sure thing that a lot of paperwork and procedures went on before the agency's representative handed you a check. Every agency is constrained by a number of rules and laws about how they come into play and how they can be activated.

Local Assistance

Local assistance includes neighborhood, city, county, and (in some cases) regional agencies. The most common venues for local governmental assistance are law enforcement agencies, fire departments, and utilities. Virtually every municipality has public safety and public works laws that require the municipality to provide for the public safety and utilities: water, power, roads, basic transportation, and fire/medical/law enforcement. Most of these services are supplemented by county and/or state services under state law.

Local NGO assistance is available from churches, charitable organizations, and local groups. Most churches have national or international aid organizations that respond locally to provide a wide range of services including food, shelter, childcare, cleanup, temporary repairs, reconstruction, counseling, and translation services. Some of the better-known organizations include the UJA Federations of North America, the Catholic Charities USA Disaster Response, the Lutheran Disaster Response, and the Southern Baptist Disaster Relief, but there are dozens of others as well as informal local church efforts.

Another example of NGO assistance is the Amateur Radio Emergency Services (ARES). Amateur radio operators in ARES provide communications services in disaster areas where normal communications channels may be damaged or limited because of demand. ARES volunteers donate their time and equipment to provide volunteer communications services to governmental and NGO agencies at all levels. For more information, you can contact the local ARES coordinator through the American Radio Relay League (ARRL) at www.arrl.org.

Other groups that volunteer at the local level include the Society of St. Vincent de Paul and the Salvation Army, both of which provide food, shelter, clothing, and other assistance for affected people. You can also expect local food banks and emergency shelters to be in full operation. In addition, local organizations may not coordinate with anyone except local authorities, such as people who set up in a parking lot and give out free water or medical care. Some companies also have their own disaster relief to deliver water or services.

The Red Cross works at all levels: local, intrastate regional, statewide, interstate regional, national, and international. They work across all these levels simultaneously to coordinate their own efforts as well as to interface with government efforts.

A great deal of individual disaster recovery is a function of insurance. Individuals purchase the insurance and the municipality regulates the insurance that can be sold, to make sure that it covers the disaster and the insurance company can cover the claims. (More on this later.)

Local disaster aid is going to be the first to get to the scene after a disaster. They're right there, after all, and they don't have as far to go. Local agencies, particularly NGOs and volunteer organizations, also aren't bound by nearly as many rules. If you're running a local food bank and you want to give away locally supplied food to someone who shows up looking hungry, you just do. A state or federal agency that has brought food in from somewhere is likely to have a lot of rules about when they can release the food and under what circumstances.

State Assistance

The general disaster response process is that the local government and its emergency services respond to a situation. If the disaster overwhelms the local ability to respond, the city or county will call in neighboring city or country resources. If this doesn't work, they'll call for state assistance.

At the state level, communities can get assistance from the state's National Guard and various state agencies. This response varies from state to state and is strongly dependent on both the type of disaster and the state's particular disaster regulations. In addition, if the disaster is affecting several counties or a region, the state provides coordination of relief efforts throughout the area.

Federal Assistance

If the state itself is overwhelmed in its disaster response efforts, the governor can make an official request for federal emergency assistance. The Federal Emergency Management Agency (FEMA) is the primary federal agency for disaster response and management, responsible for evaluating requests for disaster aid. After analyzing the situation, FEMA makes a recommendation to the president about what should be done, whereupon the president approves or denies the request.

Based on FEMA's recommendations, the president can declare a major disaster or an emergency. A major disaster declaration says that this is an event of national significance that necessitates federal aid. At that point, federal—personnel, equipment, and/or cash—assets are allocated to that disaster. In addition, a declaration of a major disaster also opens the door for long-term federal recovery programs. A declaration of an emergency is more immediate and tends to just provide funding to deal with the specific emergency need or to mitigate conditions that could result in a bigger disaster.

Initial federal funding for disasters comes from the President's Disaster Relief Fund, which is managed by FEMA and other federal agencies. FEMA's initial response is to provide search and rescue, electrical power, food, water, shelter, and other basic human needs. These resources can't be delivered until they're approved by the state, but, when accepted, federal agencies are allowed to provide assistance to the state as a whole, as well as individual assistance. Equipment and resources are done government to government, but things such as medical services or a FEMA check for a roof are individual.

As part of the recovery process, FEMA's Individuals and Households Program (IHP) is designed to provide various kinds of assistance to people in a disaster area. There are specific guidelines to qualify, but this program is basically for covering uninsured losses for individuals and homeowners. Assistance can include temporary housing; money for repairs; money for disaster-related medical, dental, and funeral costs; basic clothing, household, furniture, tools, and educational materials; and a number of other expenses. Unlike insurance, the IHP isn't designed to replace or repair everything you had to the condition it was prior to the disaster, but it will help you get back on your feet.

There are a number of additional disaster programs, including agricultural aid, financial and business loan programs, crisis counseling, unemployment assistance, hazard mitigation, and even tax and legal services. Check the FEMA website for more information.

The Disaster Relief Process

Federal disaster relief is generally coordinated under the provisions of the Robert Stafford Act of 1988, which says who can do what and under what circumstances. The process of getting federal disaster aid to a state is really like a game of ping-pong—and the state always serves.

A state government requests a declaration of disaster from the federal government. The ball goes over the net to the executive branch of the federal government, to an agency or official who's usually one step below cabinet level (which one depends on the nature of the disaster). The cabinet and the president then determine that there's a need for a national response because it's too big for a local response. Ninety nine times out of 100, it's a sure thing, but the law requires an assessment of the situation. The ball is returned to the executive branch at the state level: there's a presidential declaration of disaster and an offer to the state of federal assistance.

To return the ball at this point, the state now has to officially say, "Yes, we'll take your assistance." (This is actually a constitutional provision, because the states are autonomous and the federal government isn't supposed to just waltz in uninvited.) At this point, the feds hit the ball back and say, "Here come all the resources." To close the game, the state hits the ball back to the feds and says, "We're done, we'll take it from here, thanks!" at which point, the federal government takes off.

If the ball gets dropped at any point, bad things can happen. If a governor doesn't request a disaster, the federal government won't usually come in. (There are exceptions to this, but generally not.) And if the ball doesn't get returned at any point, the process stalls out. For example, the aid for Katrina didn't arrive in Louisiana because the ball didn't get returned at one point, so the aid couldn't be brought into the state.

States can get pre-approval for disaster relief if they know there's going to be a disaster; for example, if a hurricane or a blizzard is about to strike, the state may make a request in advance of the actual disaster. This can speed up the disaster-relief process.

In addition, it's possible for the state to send the final volley back too early. As soon as the state says, "Time to go," the federal teams pack up and leave. (State's rights coming into play again.) If the state was too early, they might need to say, "Wait, wait, please come back, we're not quite done after all!" and the feds will return.

The Disaster Recovery Center

Although FEMA is the best known and primary federal disaster agency, many other federal agencies and organizations can take part in disaster relief efforts. They all coordinate their efforts with FEMA.

A disaster recovery center is the place in the community where all the agencies and organizations tend to set up under one roof, although it may be outside the community, too. Wal-Mart, K-Mart, Target, and other big-box stores have big parking lots and they're well-known. There's enough space for everything and everyone knows where it is. It's good business for the stores, too, because it generates local good-will. Major supermarket chains do this in the Midwest. In some communities where there are lots of disasters and there are a lot of big-box stores, there's actually a competition to have a disaster recovery center on your site. (You can be a mom-and-pop store or a restaurant and get your name "on the list"; all you need to do is get in on the local disaster planning. Also, because you're part of the disaster plan, you're first in line for resources and recovery funds.) Perhaps most importantly, experience shows that being a business that volunteers time and energy to assist people during a disaster generates a lot of goodwill.

> **Storm Warnings**
>
> Volunteers can and do show up seemingly out of nowhere. They're known in the disaster business as SUVs: Spontaneous Uninvited Volunteers. These are people who want to make a contribution to the community and they will explain themselves by saying, "I'm here to help my neighbors."

You Paid for It, Now They Should Pay You

Many people may think that insurance companies are trying to hang on to each and every dollar they can, but insurance adjusters in a disaster are likely to be pretty reasonable people and you stand a good chance of getting paid off appropriately if you take a few steps beforehand.

Documenting for Insurance Claims

Regardless of the nature of your claim, you need to be able to show proof of your losses. The first thing you need to do is show the insurance company that you have coverage. As part of your planning process, make sure you have a copy of your complete insurance policy, including any riders and certificates of additional coverages.

The next thing you need to do is document the loss itself. The best way to do this is with a combination of a physical inventory of possessions and photos or video. Having a complete inventory and pictures or videos is an essential part of making a strong claim.

Almost half of all Americans don't have an inventory of their goods and possessions. Start by listing your possessions. Include the following information:

◆ Where and when you bought it

◆ The price

◆ The make, model, and serial number if available

◆ A description of the item

◆ The price

◆ Service contracts and extended warranties

Attach to this any sales receipts or the paperwork for service contracts and extended warranties. For big-ticket items, such as jewelry, art, collections, and so on, you'll want to have appraisals, proof of ownership, and photographs. Also make a point of checking with your insurance agent to ensure you have adequate coverage for these. In some cases, you'll need to get riders.

Be sure to list cars, furniture, major appliances, computers, cameras, stereos and electronics equipment, tools, and books. Don't bother taking individual inventory of generic items such as clothing, shoes, linens, CDs or DVDs, general books, and kitchenware. Just provide a count of how many of each type of item (but do make a special mention of any particularly valuable items).

It's probably going to be easiest to create the inventory list on your computer in a spreadsheet. Many personal finance and accounting programs also have options for creating an inventory of possessions.

Taking an inventory can be an enormous undertaking, particularly if you're starting this when you've been living in a house for years, but even an incomplete inventory is better than no inventory at all.

Life Preservers

Your insurance agent can probably give you a kit for preparing a home inventory with sample forms and suggestions for how to do it. It's in the insurance company's best interest for you to have a complete home inventory because then they don't have to argue with you about what you owned.

In addition to the list of possessions, you should also have photos or a videotape of the contents of your house. Take a video camera and go through each room of the house. Video all the possessions: open the cupboards and drawers and take things out; pan slowly across the books, CDs, and DVDs; video the appliances and furniture; explore the contents of all the closets; and open boxes in the garage. This not only shows the possessions themselves, but can also give an idea of their relative condition.

When you're done, save the physical inventory (in a printed format as well as on a CD), the receipts and paperwork, and the photos or videotape in your safe-deposit box or with someone outside the area. Keep a copy of the inventory and the photos or videotape with the other important papers in your Go-Pak. Between the two, you'll be sure to have proof of your possessions to show the insurance company.

The same thing goes for your house: take photos or videos of the house itself, inside and out, before disaster strikes. Have any documentation that shows repairs or improvements. If you made special preparations prior to a disaster (such as boarding up windows, cleating the roof, and so on), photograph that as well.

The rules for documenting a homeowner's claim apply to your business as well. In addition to the possessions, furnishings, and office space, be sure to document supplies and inventory on hand. As part of your day-to-day business operations, you should have a way to back up essential data such as accounting records and customer information at a secure off-site location.

When a disaster strikes, insurance companies will send in response teams to deal with claims. The insurance companies want to settle claims quickly because it's good for their business. If they're in a disaster location fast and first, they can pick up additional customers. On the other hand, if they're slow to respond and difficult to deal with, they're going to lose customers. If you make it easy for them to approve your claim, they'll be able to get to the next person. They aren't going to make a fuss out of something that's well documented. The claims that are going to be held up are the ones the insurance company thinks it has a fair chance of contesting.

Mitigating Losses to Maximize Payouts

Suppose you've been through a hurricane and the following things happened to your house:

◆ The hurricane blew the roof off your house. Rain, driven by the hurricane, got into the house and caused damage to the second floor.

◆ A day or two later, what was left of the hurricane has dissipated, so the hurricane is officially declared "over." Nevertheless, there's still bad weather, so a rainstorm (which is officially not part of the now-dead hurricane) comes by and soaks your second floor again, this time soaking everything that was still reasonably undamaged.

◆ The day after that, a levee nearby, weakened because of all the rainwater and the pounding of the wind, bursts and sends a surge of water that now floods your previously dry, intact first floor.

It's now time for you to make a claim for damages. To most people, all of this seems as though it's one big claim: the hurricane came through and blew off the roof, it rained all over you a couple times, and then a levee burst because of the extra rainwater and you're soaked from top to bottom. But this isn't the way the insurance company is going to view it.

Each incident is going to be treated separately because it potentially falls under a separate part of your policy, with separate actuarial risks and separate reimbursement options. The initial loss of the roof is part of the hurricane. The rain that accompanied it is most likely going to be classed as "wind-driven water," which may or may not be covered, so you could conceivably be covered for the loss of the roof itself but not for the damage caused by the initial soaking. If it's part of the hurricane damage, though, it may all be treated as part of the hurricane. To complicate things, homeowners' policies in areas that get a lot of wind frequently have wind damage as a separate policy rider, whereas in other places, it's just part of your general coverage.

The second rainstorm would count as a separate claim: a rainstorm came along and soaked your house to the point that it caused damages. This might well be covered under your general homeowner's coverage, but it might not. For example, the insurance company may want to know if you did anything to mitigate the fact that you didn't have a roof, leaving your house open for just this kind of damage. (Yes, this may sound picky, but it's the way this all works.) It's likely that you could show that you didn't have any way of mitigating the damage because there were no blue tarps available for 30 miles in any direction, no open roads, and everyone else in your part of the city had suffered similar damage, but it's something you may be required to prove.

Showing that you've done something to mitigate damages is key to getting favorable treatment. You can't just wait around for weeks without doing something to fix your roof; you have to try to find blue tarps or sheets of plywood and put them down so the next time it rains, the top half of your house isn't wide open.

Finally, the flood from the levee could be covered or not for half a dozen reasons. The answer hinges on certain questions such as—is this caused by the hurricane, are there other proximate causes (such as the break was caused by someone's negligent or willful actions), do you have federal flood insurance, and do you need federal flood insurance? If you live in an area that's outside a designated flood zone, you may not be required to have flood insurance to be covered.

Any number of factors go into the determination of an insurance claim, including the following.

- Federal laws

- State laws

- Local laws

- The type and quantity of insurance coverage you have

- The identified conditions where you are

- If you have federal flood insurance

- If this was an officially declared major disaster

- If you attempted to mitigate the damages in between incidents (for example, did you try to put a temporary roof in place before the second rainstorm hit?)

- If you've hit the policy coverage limits for this claim or this type of claim

The bottom line is that when it comes to insurance coverage, *it all depends.* You should be aware of the general coverages you have. It's also a good idea to review your coverage every 12 to 24 months with your insurance agent and see what coverages you need to increase or add.

When Friendly Insurance Isn't So Friendly

Suppose you think that the insurance adjuster you're dealing with is being difficult, isn't very good, or is just simply a jerk. Most of them are pretty reasonable people, particularly in a disaster situation, but you may run into someone who you think isn't doing their job the way it should be done.

The first step when you're dealing with an insurance adjuster who isn't giving what you want is to determine if they even can. Ask the insurance adjuster, "Do you have the power to say 'yes' to my claim?" If they can't, you know that this person isn't going to be able to help even if they want to. At that point, you can ask to speak to someone who can approve your request. They may not, either, but at least you're a step closer.

If this person can approve your claim, you next need to ask, "Do you have the power to say 'no' to my claim?" This may sound like an odd question, but adjusters in a disaster situation frequently are authorized to say "yes" but not "no." A "no" requires them to kick things up to a review committee who may or may not say "no." So if they can say "yes" to your request (but they could) but they can't say "no" (but they aren't), it means that they want to say "no" and they're trying to wear you down without saying "no." You may be able to work with this and see if you have room to negotiate the size of your claim.

However, if you just can't get satisfaction out of the insurance adjuster, you may be able to hire a certified professional public adjuster (CPPA) and get them to render their own opinion. CPPAs are professional adjusters who know how to interpret insurance coverages and also how to assess losses. They're also familiar with all aspects of the claims process and may have inside knowledge about presenting a claim that you don't.

You've probably already met at least one CPPA in your disaster claim process: the person you haven't been happy with representing the insurance company. When a disaster strikes, the insurance companies hire lots of CPPAs in the area to take care of all the extra claims they have to process. They're already there and can do a good job for the companies. The thing is, the insurance companies are likely to hire the CPPAs who are likeliest to be sympathetic to the insurance companies' point of view. Sure, the companies want to look good and settle claims quickly, but they don't want to be profligate. CPPAs who lean heavily to the customers' interests may be last on the list to get hired for disaster claims—and that's a great advantage for you.

Any available CPPAs in the area are, by process of elimination, not friendly to the insurance companies, so they're going to be good for you. Now you have someone on your side who's certified to identify the hazard, the cause of the loss, and the value. CPPAs don't adjudicate if you have coverage—you still have to wrestle that one out with your insurance carrier—but they do examine your policy carefully to identify every possible option for reimbursement. They also identify the causes of the loss and state a value for the loss, which may well be higher than the insurance company has stated.

CPPAs cost about the same as lawyers, with typical expenses running from $150 to $300 or on a contingency basis, taken as a percentage of the settlement with the insurance company. It's also possible, however, that you can charge the costs for the CPPA back to the insurance company as part of the settlement, which gives the insurer an incentive to realign their adjustment toward that of the CPPA.

On top of everything else, there's another reason the insurance company is going to be inclined to settle favorably at this point: people who know the system well enough to get a CPPA are one step away from suing the company. In fact, most CPPAs work with lawyers specializing in suing insurance companies. If too many people sue an insurance company over their failure to pay claims in a disaster, the chances are good that this can become a class action suit, which can cost the company millions in damages and in lost business.

Chances are good that, if you have your paperwork in order and you're not asking for the moon, you have a fair chance of getting a settlement that's at least close to what you want.

Businesses have special business issues to deal with. In addition to the normal fire/theft/general loss business insurance, most businesses also need liability insurance. In addition, there are other types of insurance to consider, such as errors and omissions insurance, which you can think of as malpractice insurance for whatever it is that the business does, and business interruption insurance, which insures against losses of income in a disaster.

The Least You Need to Know

- ◆ The most expensive part of a disaster is the recovery phase, particularly when there has been damage to roads, public facilities, and the infrastructure.

- ◆ Although many agencies and organizations are involved in a recovery, the local disaster recovery center is the place to find financial aid and assistance sorting through the labyrinth of those offering aid.

- ◆ Having your policy numbers and copies of the policies will help establish the type and limits of your coverage.

- ◆ When you're making an insurance claim, a picture's worth $1,000—or maybe a lot more.

- ◆ Insurance agencies have many reasons to approve claims after a disaster. All it takes to get a "yes" is to make their job easy.

Chapter 17

If You Failed to Plan

In This Chapter

- ◆ What to do if you failed to plan
- ◆ Understanding strategic thinking
- ◆ Gathering the right kind of information
- ◆ Managing your message to set expectations

The adage "failing to plan is planning to fail" has become the mantra for every manner of continuity and crisis. It's certainly better to be prepared for every possible hazard with a contingency plan based on a business triage/resource allocation model. But even so, your business may not have a plan in place to deal with a crisis.

In this chapter, we show you what to do for your business if you bought this book and forgot to read it or you bought this book after the disaster struck. In other words, in this chapter, we help you survive and thrive even when you are not prepared.

Businesses That Fail to Plan

Although it's always better to plan for success and prepare for crises, your business can still be successful even if you haven't planned. And even when you've planned for reasonable hazards, such as floods, fires, and earthquakes, you may not be able to handle an event that is so novel that it simply couldn't be anticipated, such as a volcano erupting.

Even without a plan, following these six steps will greatly improve your chances of weathering a disaster successfully and with your business intact:

1. Don't panic.
2. Use common sense.
3. Work smarter, not harder.
4. Don't wander.
5. Seek help.
6. Mitigate outrage.

Don't Panic

Not panicking is far easier said than done. Nevertheless, fear and panic can cause even the best leaders to act for the sake of acting. It's possible that doing "something" may be inspiring, courageous, and even the right action, but it's also possible that this "something" may cause financial loss, physical damage, or even injuries.

Strategic Thinking

Strategic thinking is a skill you might need to practice. Most people tend to think *tactically:* figuring out what will get them from A to B. Strategic thinkers think in terms of what will get them from A to Z. When dealing with a disaster, you need to think strategically.

def•i•ni•tion

A **tactic** is a single maneuver or exercise that accomplishes an objective. A **strategy** is a plan or series of tactics to accomplish an overall goal.

Strategic thinking gives you certain advantages. First, you know what you're trying to accomplish. Second, with a plan in place, the tactics for accomplishing the plan can start falling into place. Third, it gives a sense of confidence because you know what you're trying to do.

Brown's Law states "If the patient is still breathing, you have time to think." In the midst of a crisis, as long as there isn't an immediate and direct threat to life, limb, or property, you need to stop and think. You need to know what's happening, what the immediate threats are, and what risks you're likely to face.

When you're facing a disaster, the first thing you need to do is stop and take a breath. It's calming and forces you to stop for a second. The second thing is to take your own pulse. Again, this is calming and refocuses you. (And if your heart's really pounding, as is usually the case when facing a crisis, this will help you realize that you're feeling hyped up.)

You're a bit calmer now, so the third thing you need to do is gather information and then think about it. When you've thought about the things you need to do, the fourth thing is to do something. The fifth thing is to wait long enough for something to have happened. When you've done something tactical, you need to wait long enough to make sure it's had the desired effect. If you do something, then immediately do something to "correct" it, you may make things worse rather than better.

The final step is to repeat the process: breathe, check your pulse, gather information, think, do something, and see how it worked. Short of someone actually dying or about to go off a cliff right in front of you, there are very few things in life that you can't use this process for.

Gather Information

Step three of the process is to gather information. In fact, crisis management involves gathering information on five levels:

- Macro-strategic: The view from 10,000 feet
- Strategic: The view from the rooftop
- Operational: The view from the street
- Tactical: The view from the trenches
- Micro-tactical: The view through the eyes of the employees

The type of information you're going to gather depends on the role you fill in the organization. After you gather this information, you can assess the full extent of the disaster on your business.

The macro-strategic level is the CEO, the business owner, or whoever's in charge. The boss needs to know what's going on so that he or she can formulate a strategic plan. One notch below, at the strategic level, are people who know the strategic plan but they don't have the power to change it. Their job is to take the plan and break it into individual parts and then hand it down.

The operational level is middle managers. This could just as easily be identified as the "micro-strategic/macro-tactical" level. They're the universal translators. They're not data filters; they're facilitators. This level takes the strategic information they've received from above and feeds it to the operations people below them. They also sort and collate the information they're receiving from below. The middle managers know what's going on at the bottom level but they're just monitoring it. One other thing they can do is flag what's going on. If something goes wrong, it's not their job to fix it. It's either the person below them who needs to fix the operations problem or the person one level above them if it's a case of fixing the strategic plan. And if it's all green, the middle managers can flag things as working so the person above them can look at everything and know that it's all good.

The tactical level is the group supervisors, the operational heads, and the shop foremen. These people are purely tactical. They monitor what's going on at the bottom level and report this up the chain, good or bad, to the middle managers. And at the bottom, the micro-tactical level, are the people who are actually in the trenches dealing with customers, working machines on the floor, or programming.

In general, everyone knows the plan one level up and one level down from wherever they are. They shouldn't know directly what's going on farther away than that, though. The people at the top don't need to know the details from the shop floor, and the workers at the bottom of the pyramid don't need to know what's going on overall. They just need to know what the next thing they need to accomplish is.

A common mistake during disasters (as well as a common business mistake) is for the senior people to micromanage. This happens when the people at the top get scared and don't trust the people at the bottom.

It's not the role of the senior people to know what's going on at the bottom. Instead, the very best way to manage during a disaster is to not change your operational response at all. It should be "business as usual" as much as possible.

For example, the crews in the engine room of a nuclear sub don't need to know that there's a hurricane blowing overhead. There's no reason to restrict the information, but neither is it useful to them and it won't help them accomplish their tasks. The captain needs to worry about that, but he doesn't need to worry about how something loose got lashed down in the engine room.

Micromanaging during a disaster frequently manifests as everyone tries to force themselves into the bottom level. You'll often hear the phrase, "I want to make sure the job is done right" (or "*still* done right"). This is invariably said by people who've never gotten comfortable with the fact that their jobs are a couple layers above where they used to be. There's a difference between the CEO who walks the floor and greets the workers and the one who sticks his nose into the bins. The micromanaging is a trust issue: the manager hasn't learned to delegate.

Use Common Sense

Gathering information is vital, but you should use common sense, too. If there is something obvious you can do that will stem the deepening of a crisis, do it! Common sense can also help you another way: the rules of business and the laws of nature don't change in a crisis, so if the action you're thinking of taking wouldn't make sense when there's no crisis, it's unlikely to be productive during a crisis.

Don't overthink things. Too much observation can become analysis paralysis. You keep looking at the issues and never come to conclusions, allowing the effects of the disaster to worsen. You might also be trying for the "perfect" answer. Don't look for perfect; "good" is just fine. In this case, "close enough" is not just for hand grenades and horseshoes.

Framing bias is where your perceptions have been colored before you have an opportunity to analyze something for yourself. Suppose you're an ER doc, and a patient well known to the ER department is in Room 3. The nurse hands you a chart and says, "Drug-seeking behavior ...," and you've just met framing bias. The information from the framing bias may be true or false, but the important thing is that it has a high potential to cause you to exclude information that's necessary for coming to the right conclusions. (The framing bias in this example, by the way, could conceivably cause the patient great harm, because the doctor would be more likely to dismiss complaints about pain as part of the overall drug-seeking behavior.)

Framing bias is also the tendency to view a problem differently because of the way in which it is stated. If you said, "The business has a 20 percent chance of failing as a result of this disaster," it's technically the same thing as saying "The business has an 80 percent chance of surviving this disaster." Nevertheless, focusing on failure rather than success in this example makes the statements appear to be saying different things.

Make sure you ask yourself the right questions. If you've had a flood, the right question may not be "How can I get the business open?" or even "How can I provide services to my customers?" but instead "How can I get my employees to come to work?" As you're looking at the problems, see if you can change the way they're stated to show the "other side," or at least make them more neutral. Turn things around so that you're considering them from another side: "I'm an employee; why would I want to go to work right now?"

One powerful technique is to think about problems without using the word "but." For example, instead of saying, "I want to get the employees to come to work, but the streets are muddy *and* hard to drive on," say, "I want to get the employees to come to work and the streets are muddy and hard to drive on." The first sentence is a complete thought that really doesn't take you anywhere, while the second sentence is framed as though you're setting up the parameters of a problem. With the second sentence you might then say, "Okay, if the streets are hard to drive on, how else can I get the employees to come to work?"

Work Smarter, Not Harder

Good leaders know that the smartest way to work is to rely on past successes as a blueprint for future success. This requires you to use pattern recognition, a form of thinking often mislabeled as "gut" or "instinct."

Pattern Recognition

Pattern recognition is the act of matching the current situation at all five levels of observation with prior situations. The best matches predict how well something's going to work as the result of repeating prior actions. If a prior action led to success, repeating that action should result in success now. Conversely, if a prior action led to failure, then that is an action to avoid.

Ironically, many managers, particularly those with technical rather than business backgrounds, don't think of things this way. It is not uncommon for them to keep trying the same thing in hopes of "getting it right." However, the penalties for failure are usually much higher during a disaster, so before you repeat any action, make sure you're doing this because it actually worked, not because you think it *should have worked* last time.

Heuristic Deduction

Heuristic deduction is a fancy name for the act of mentally walking a mile in the shoes of another. Through the techniques of heuristic deduction you can infer the needs and even responses of other people based on your own needs and responses when in the same situation. To do this, you need to have empathy and be able to recognize your personal biases, but after you master these, heuristic deduction becomes almost automatic when you make plans.

You actually already do a lot of this, possibly without realizing it. If you say, "How would the customers respond if we did this?" or try to imagine how your employees or family would react to various kinds of news, that's heuristic deduction.

One principle of heuristic deduction is that you can make predictions about the behavior of a whole group of people if you can learn to act and react similar to the people in the group. Some professions use this technique all the time. Ad agencies and marketing departments work this way: their people can assume the traits of the market they're after and say things such as "If I were a 35-year-old housewife, here's how I'd react to this approach." By taking on personas, chameleon-like, they can actually test what they're doing before sending ads out to focus groups, making the process much more effective.

Don't Wander

Wandering is losing sight of your goals and objectives. You need to remain focused and work toward the desired ends. This process is known as triage. You determine exactly what your priorities are in an emergency. Business triage is no different: you need to prioritize your goals, outcomes, and processes, and then allocate the resources objectively based on the relative importance for those goals and outcomes.

Triage is not simply sorting the most important project, business goal, or critically ill patient to the front of the line. Triage is determining what resources are available and how those resources can serve the largest number of goals or people at any given time. Triage is a continuous, repeating process. In business, this means you must constantly reassess the resources available at hand as they are used up and replenished. You must also reassess the needs and goals of the company on a minute-by-minute, hour-by-hour basis as the situation demands.

The first step in business triage is to identify and categorize the desired outcomes, goals, or objectives:

◆ Critical/essential outcomes are those that must occur to meet your overall mission.

◆ Urgent/important outcomes are those that facilitate the overall mission but are not absolutely essential to it.

◆ Supportive/optional outcomes are those that facilitate the mission but aren't necessary to it.

After you've classified the outcomes, identify and categorize the processes that result in the desired outcomes in a similar fashion:

◆ Critical/essential processes are those that must occur to meet your overall mission.

◆ Urgent/important processes are those that facilitate the overall mission but are not absolutely essential to it.

◆ Supportive/optional processes are those that facilitate the mission but aren't necessary to it.

With your lists of outcomes and processes, you're ready to assign resources and prioritize. The critical/essential outcomes are fully supported first with all available resources. After that, the urgent/ important outcomes are supported, and, if there's anything left over, the supportive/optional outcomes. If there are several critical/essential outcomes, you can assign priorities to them or assign a different group of people to each one. This process ensures that the organization does not wander from the path to reach the desired outcomes.

The most important thing for you to understand is that when a business faces a disaster it won't be able to accomplish all its goals and objectives until more resources are brought to bear. If those resources are not available, then you must identify the acceptable losses, dispassionately and with the same logical approach the business uses when choosing a vendor or investing in a new project.

Short-Term vs. Long-Term

When you're making decisions about your priorities, avoid short-term tactical gains that have long-term strategic costs. For example, if your job suddenly went from full-time to part-time, and you got the maxi- mum cash advance from your credit cards to meet immediate expenses, you'd have a short-term tactical gain (the bills got paid) but a long-term strategic cost (repaying all the high-interest cash advances).

On the other hand, a short-term tactical loss can also result in a long- term strategic gain. Several years ago when New York City suffered yet another blackout in mid-summer, Arnie, who owned a small con- venience store and ice creamery, faced a business triage decision. His customers had already been in to purchase what he had on hand, but with the power out he had 10 flavors of ice cream in the cabinet that would soon melt. At 5 gallons per flavor, there was slightly less than 50 gallons of ice cream up front and another 100 gallons of ice cream in the back.

Arnie knew that he had a disaster on his hands: his needs (refrigerator) exceeded his resources (electricity). Arnie needed to decide where he could focus his efforts and his remaining resources so that his business would in fact reopen when the power came back on. He also needed to plan for as short a recovery as possible. It takes a lot of effort to get rid of 150 gallons of ice cream, and a lot of dumpster space. The cleanup would be awful and, if the power was out for too long, his store would be filled with the stench of sour milk and rotting ice cream.

Arnie decided to give the ice cream away. He wouldn't lose anything, because the product would spoil before the refrigeration returned. So he put a small handmade sign in the window that said "Free Ice Cream."

In no time, he had a line. He was giving away the ice cream, but what to hold it in? Ice cream cones! The cones were actually cheaper than Styrofoam cups, and Styrofoam has an unlimited shelf life. Would the ice cream cones go bad during the blackout? No, but you can't give people ice cream in their hand, and the small loss in the cost of ice cream cones was less than the larger cost of Styrofoam cups.

To Arnie's amazement, many people tried to pay him for the ice cream. Wanting to get rid of it as quickly as possible before it all went bad, he simply refused. To his greater amazement, people began to buy other items in the store, including many items that he wouldn't have been able to sell right then because he hadn't had many customers before. Before he had given away all the ice cream, Arnie's store shelves were bare. His acceptable loss, the ice cream, had gained him an unexpected profit.

When the power came back on, Arnie was re-supplied with ice cream and merchandise. To his surprise, he also saw a tremendous increase in business. People didn't just come back because he had given away ice cream. They returned because they felt that Arnie cared about them. He had taken a tough decision and turned it into a benefit for those around him.

Bull's-Eyes, Budgets, and Boundaries

As you set your priorities, you need to consider bull's-eyes, budgets, and boundaries. Bull's-eyes are the individual tactical goals that the workers at the micro-tactical level are working on. You need to prioritize individual bull's-eyes so that you can set their budget, the number of resources available for them. After you set the bull's-eyes and their budgets, specify the boundaries, which are the time, resource, and working environment limits for the workers.

Seek Help

Don't go it alone. There is strength (and creativity) in numbers. Although this should go without saying, it is essential that when a disaster strikes, you seek advice from experts as to the best ways to manage the crisis. This may be your first flood, but there are people who deal with them all the time.

You also should look for help and advice on communicating about the situation and the risks to your employees, your customers, and the general community, to keep a crisis from becoming a public relations catastrophe.

Mitigate Outrage

Mitigating outrage is essential to keeping your employees and your customers happy. The secret to this is to set expectations appropriately so you're sure to meet them.

As you saw in Chapter 1, the equation for outrage is the difference between your experience and your expectations. You want to mitigate the potential for outrage as much as possible. You can't affect someone's perception of an experience, but you can definitely set their expectations. In this way, a business can mitigate or even eliminate outrage and thereby change the way the customers feel about you.

The expectation that you create for others is based on what you put out there. For example, you take a seat in an ER and you're told that you'll be seen by the doctor in 15 minutes. Well and good, but it actually takes about 20 minutes to do the paperwork. The result is that you feel as though they're slow or inefficient. Your expectations were set and then not met.

On the other hand, Disney sets and handles their customers' expectations brilliantly. Suppose you're in a long line for Space Mountain and the sign said that the wait was 75 minutes. If it only took 60 minutes to get on the ride, you'd be thinking "I got 15 minutes of my vacation back." But if the sign said it was a 60-minute wait and it then took 62 minutes, you'd be unhappy, even though you'd lost no appreciable amount of time (and it's a 3-minute ride in any case). The key is that

Disney has created the expectation for you. That's controlling the outrage: you set up an expectation that you're sure you can meet.

On the off chance that Space Mountain breaks, there'll be an announcement that the ride is temporarily unavailable. Some people will leave the line; others will stay. Why? Because with all those other folks leaving, they'll be able to get on that much sooner. If the ride remains down, the announcement is repeated every 5 minutes. More people will leave the line, and still more. If it becomes clear that the ride is not going to be up for the rest of the day, Disney employees are sent out to personally apologize to the remaining people in line. Most of the time, the people are okay, but if they aren't, the employees have fast passes to every Disney ride in their pockets, so you can get on another ride without having to wait. And chances are that you'll be back the next day, waiting another 60 minutes for a 3-minute ride, and you'll be really, really happy because you got a free ride without having to wait in line, despite having waited in line for a ride that was broken.

What you tell your employees and customers is important. As Winston Churchill said, "If you mean to profit, learn to please." You need to show that you know what effects the disaster is having on your business and how this will affect the workplace. In addition, you need to set their expectations for what you will do during the crisis and then come in bigger and ahead of schedule. The old business rule of "Underpromise and over-deliver" is never truer than during a disaster.

The Least You Need to Know

- Failing to plan is *not* planning to fail.
- One of the best ways to inspire people and instill confidence is to not panic in the face of disaster, but to act with calm deliberation.
- In a disaster, the true leader is the one who remains a leader. The bad leader becomes a worker bee.
- The best disaster is one in which the worker bees never notice that something has changed.
- Manage expectations and you will be a hero.

Chapter 18

Putting It All Together

In This Chapter

- ◆ Disasters will happen sooner or later
- ◆ Preparing for disaster
- ◆ Surviving the disaster when it happens
- ◆ Learning from your experience and improving your plan

Disasters can strike anywhere at any time. They can be frightening and disorienting, but through preparation, planning, and practice, you can survive and even thrive in the face of adversity and disaster. In this final chapter, we review the essential skills and principles you will need to have mastery over adversity.

Practice the Way You Will Play

Throughout this book, there have been two main messages. First, *a disaster will happen to you*. All disasters by definition are situations in which your needs exceed the resources you have available, so the nature of the disaster that will occur can be almost anything. It can be a purely natural disaster, such as an earthquake, a hurricane, or a flood. It can be a manmade disaster, such as a toxic spill or an explosion of some kind. It can even be a social disaster, such as a major economic collapse. There's no telling, but it's certain that there will be a disaster in your future.

Second, *no matter what the disaster, there's a great deal you can do to miti- gate the effects.* By focusing on supporting and maintaining the processes of your life and business that are most important, you can keep the most important parts of your life going while you're looking for ways to get more resources and restore the remaining processes. The very best plans are the ones that let you continue with your life without any noticeable interruptions. This book has shown you how to get ready for a disaster. The steps are actually very straightforward.

Learn everything you can about disaster preparedness. There's an awful lot of information available about disaster preparedness. Appendix A has a wide range of websites and other resources that can give you general disaster information, making and buying Go-Paks, dealing with children and people with special needs, how to get financial aid, business information, weather and disaster alerts, community and faith- based resources, and dealing with animals in disasters. Pick an area or two of particular interest and start there. You'll be sure to find some- thing useful and interesting almost immediately.

Check with your local emergency operations center. Every state, county, and major metro area in the country has an office of emergency management by one name or another. These offices are responsible for coordinating emergency operations in their area. You can obtain lists of emergency contacts and shelters, recommended evacuation routes, maps, and assistance creating disaster plans.

Prepare a Go-Pak for you and everyone in your tribe. You and everyone in your tribe need to have Go-Paks. A Go-Pak has everything you need to keep you alive for the first 72 hours, by which time you should be able to connect with additional resources from external disas- ter relief. You should each have a Go-Pak at home and one for the car. Appendix B provides you with information on what to pack in a basic Go-Pak as well as suggestions for things to add for children, pets, and special needs people.

Create a disaster plan. Without a disaster plan, it's entirely possible that you may not survive a disaster. You're at the mercy of circum- stances, and there's no reason you have to be. Appendix C has forms and checklists to help you assemble the information for a disaster plan tailored to your needs.

Make copies of your important documents. It's essential that you have copies of essential documents, such as marriage certificates, medical and school records, licenses, and passports. A basic list of documents to carry with you is in Appendix B. You should also have physical or digital copies of important family photos and other records.

Check your insurance coverage. Insurance won't prevent a disaster, but it can help you recover from one. Review your insurance coverage for your home and business and make sure that everything's current. Whenever you get added coverage, be sure to update the copies of insurance information in your important documents, too.

Practice, practice, practice! A disaster plan that's never been tested is as bad as no plan at all: you don't know if what you'll do will make things better or have any effect whatsoever. In fact, an untested disaster plan can even expose you to more risk than if you did nothing at all.

Practicing your disaster plan also lets you spot little errors that can have big effects in practice: one hospital drilled their employees on the disaster plan, but because everyone was thinking that it was a drill, nobody noticed that none of the flashlights had batteries until an actual emergency happened.

Practicing your plan helps you stay calm because you know immediately what you need to do next. You don't need to look it up. You also know you can do it, because you've done it before. There's nothing new about the disaster plan to surprise you. In the middle of coping with a disaster, a surprise may quite literally be the very last thing you want.

Play the Way You Will Play

If you've been practicing the way you will play when disaster does strike, you'll be ready for it. You'll have your Go-Pak and a disaster plan. Appendix D has checklists for getting out the door safely and to a shelter or evacuation site, or for sheltering in place as the case may be. Because of the time you spent practicing your plan, you'll know what to expect at all stages of the game, your expectations are unlikely to exceed the experience, and you'll feel reasonably calm about what you need to do.

As you're going through the disaster and working your plan, be sure to make notes when you can about how the plan is working. You will observe many lessons, but they may also be lost because they're happening too fast. When something occurs to you, make a plan for how you're going to solve the problem in your disaster plan, and then do so.

When life returns to normal, assess the overall effectiveness of your disaster plan. Things may have gone smoothly enough during the disaster, but when things calm down, you may realize that things didn't work as well as they might have, or that you supported processes that weren't as important as you thought they would be. You can always change your plans for the next time, but be sure to practice the new plan to make sure the new parts work well with the rest of the plan.

Look at how well your community's disaster response worked, too. You may be able to help improve or refocus community responses. You might also be able to pool your efforts with your neighbors or friends to come up with a disaster plan that will help everybody in a neighborhood much more than each person or family could manage on their own.

Finally, your most vulnerable time for a second disaster is immediately after a first. Your supplies are depleted, the community's resilience is reduced, the infrastructure is likely to be damaged, and houses and buildings are weakened. This is why hurricanes that hit within the space of a few weeks are so much more devastating than they would be individually, or why flash flooding after a wildfire takes out so many houses. As soon as you are able, replenish your emergency supplies and your Go-Paks, refuel your car, and make repairs to your house. Update your disaster plan and have a drill shortly after the disaster when the lessons are fresh in your mind. You may need it sooner than you think.

Your ultimate goal is not merely to survive, but to thrive. Plan for your success, practice your plan, and apply the lessons you learn for the next time. If you keep focused on maintaining the processes of your life and business, you should have no trouble with both.

The Least You Need to Know

- Pack your Go-Pak now!

- Have a plan and practice it.

- The best emergency is one that no one knows has taken place.

Appendix A

Additional Resources

This appendix is a list of online and print resources to assist readers in finding additional information on a variety of disaster preparedness topics, products, checklists, and teaching tools.

Websites

This section contains URLs for websites for disaster and preparedness information. If a URL doesn't work, try googling for the site using information from the description. Frequently, a site will change its structure or domain, but the information is still available. You can also find many web pages that don't exist anymore by entering the URL in the Internet Wayback Machine at www.archive.org.

Author Websites and Resources

www.MauriceARamirez.com
Informational website for Dr. Ramirez as a professional speaker for your company or organization.

www.Disaster-Blog.com
The official blog for Dr. Maurice A. Ramirez, founding chairperson of the American Board of Disaster Medicine, and expert in the areas of disaster preparedness, disaster medicine, and human continuity.

EzineArticles.com/?expert=Maurice_Ramirez
The official online article bank for the writings of Dr. Maurice A. Ramirez. Articles downloaded from this site may be used as content for websites and print periodicals in compliance with terms provided at this website.

www.hedtke.com
The website for John V. Hedtke and JVH Communications, a private company that provides documentation, training, white papers, and other communications services to a wide variety of private, public, and government clients.

www.CompleteIdiotsGuide2DisasterPrep.com
Online link to purchase this and other disaster-related topics by these authors.

General Disaster Information

www.bt.cdc.gov
Subsection of the official site of the Centers for Disease Control and Prevention (CDC) dedicated to emergency preparedness issues. Links to detailed information about bioterrorism, disasters, mass casualties, and recent outbreaks of infectious diseases. Follow the link to the "Emergency Preparedness and You" page for recommendations about emergency planning and preparations for the home, community, and business.

www.cbsnews.com/digitaldan/disaster/disasters.shtml
Comprehensive list of links to national and international disasters sites: types of disasters, weather alerts, relief agencies, educational sites, business and technology resources, emergency management sites, and an interactive map to connect to state-based resources.

www.chefnoah.com
Independent website dedicated to family preparedness and provident living. Includes practical suggestions for families regarding emergency preparedness, food storage, cooking tips, and home safety.

www.cidrap.umn.edu
Site of the Center for Infectious Disease Research & Policy at the University of Minnesota. Provides up-to-date information on influenza, bioterrorism, and infectious disease topics.

www.deep.med.miami.edu
Website of the Center for Disaster and Extreme Event Preparedness (DEEP) at the University of Miami (Florida). The Center is involved in training, research, and services focusing on behavioral health, special needs preparedness, and epidemiologic aspects of disasters.

www.dhs.gov/xprepresp
Preparedness and response page of the official site of the Department of Homeland Security (DHS). Includes links and downloads related to disaster and preparedness-related programs, government responses, training and education, publications, laws and regulations, federal committees, and local resources.

www.disastercenter.com
Online coverage of disasters in the United States. Includes weather maps, links to disaster studies and statistics, and links to information and resources specific to each state (TV, newspaper, radio, maps, weather, and government).

www.disasterhelp.gov
Separate site of the Federal Emergency Management Agency (FEMA). Detailed information on current disasters, assistance programs, emergency resources, shelters, gathering loved ones, and links to local resources. For descriptions of types of disasters and numerous downloadable booklets and brochures, look under the "Disaster Resources" tab.

www.disaster-zone.com
Blog dedicated to information-sharing on a variety of Homeland Security and Emergency Management topics.

www.fbi.gov
Official site of the Federal Bureau of Investigation. Follow links on the home page to find information on terrorism, national security, and numerous government reports.

www.fema.gov
Official site of the Federal Emergency Management Agency (FEMA). Follow home page links to obtain information about types of disasters, how to prepare your family and your home (including pets and livestock), disaster assistance programs, business recovery, and teaching tools for parents and teachers.

www.floridadisaster.org
Website of the Florida State Emergency Response Team (SERT)
dedicated to disaster planning for the home, business, and community.
Includes tools to develop emergency plans, tips on strengthening
your home against natural disasters, Florida state emergency/disaster
resources, legislation pertaining to disasters in Florida, and resources
for educating children about emergencies and disasters.

www.hhs.gov/disasters
Official site of the U.S. Department of Health & Human Services.
Natural disasters, man-made disasters, and traumatic events are defined
and numerous educational resources for preparedness and response are
provided.

www.informedguides.com
Link to pocket resource guides. Follow the "Home" tab for publica-
tions about Home Emergencies, Homeland Security topics, and Pet
Emergency planning and management.

matrix.msu.edu/~disaster
Disaster Mitigation and Planning Assistance website, a joint venture
of the University of Michigan, Center for Great Lakes Culture, and
the California Preservation Program. Provides readers with a search-
able menu (by state) of disaster-related services, supplies, and access to
experts.

**mentalhealth.samhsa.gov/publications/allpubs/KEN-01-0096/
default.asp**
Official website of the U.S. Department of Health and Human Services
National Mental Health Information Center. Tips on disaster counseling.

www.nrcs.usda.gov
Official website of the National Resources Conservation Service of the
USDA. Provides information on soil erosion, flood control, and miti-
gation of damages by natural disasters. Follow link to the Emergency
Preparedness page for even more information.

www.pandemicflu.gov
Official website of the U.S. Government devoted to educating the public on all aspects of pandemic and avian influenza.

www.pdc.org
Website of the Pacific Disaster Center, an organization devoted to reducing disaster risks and impacts to peoples' lives and property in the Asia Pacific and Indian Ocean regions by integrating various technologies for disaster management.

preparednessandresponse.blogspot.com
Blog about disaster preparedness. Includes an extensive list of other disaster preparedness blogs and websites.

www.ready.gov
Website of the Department of Homeland Security, providing the general public with information, resources, downloadable materials, and checklists to assist families, businesses, and schools to prepare for disasters.

www.redcross.org
A wealth of information about the services, products, educational resources, online training videos, and response efforts of the American Red Cross. Read about Red Cross responses in the Disaster Online Newsroom. Also look for resources for children, including coloring books, disaster information, and educational curricula.

www.thebeehive.org/emergencies
From the public information website of the One Corporation, a global nonprofit organization focused on bringing technology and information to low-income persons on numerous topics (home, health, jobs, schools, finances). Contains written information, brief instructional videos, and checklists for preparing yourself, home, and family for emergency and disaster situations.

www.udel.edu/DRC
Informational site of the University of Delaware's Disaster Research Center. This Center conducts research on preparedness, response, and recovery by the public in relationship to disasters and crises affecting communities.

www.unisdr.org
Official website of the United Nations International Strategy for Disaster Reduction (ISDR). The ISDR's mission is to build disaster-resilient communities by promoting awareness of disaster mitigation. Provides a wealth of information about types of disasters, statistics, recovery efforts, meetings/conferences, and other educational resources from around the globe.

www.usace.army.mil
The Army Corps of Engineers website, with flood control information, local contact information, maps, and projects.

www.usbr.gov
River and dam information, water levels, water supply, and other environmental data.

www.usgs.gov/hazards
Scientific information about natural hazards including earthquakes, floods, hurricanes, landslides, volcanoes, tsunamis, and wildfires, and resources for disaster help to minimize loss of life and property from natural disasters.

Making a Go-Pak

www.1800Prepare.com
Online source for 72-hour Go-Paks.

www.beprepared.com
Online source for 72-hour emergency kits as well as food supplies, emergency communication items, and water filtration/purification supplies.

emc.ornl.gov/CSEPPweb/animals/animals_emergencies_owner.htm
Suggestions for Go-Pak contents for pets.

www.emergencylifeline.com
Emergency kits for home, business, community, and search and rescue teams.

www.fema.gov/plan/prepare/basickit.shtm
Lists of items to include in preparedness kits for home, work, and car. Follow link to "Are You Ready?" for downloadable video and checklists.

www.providentlivingcenter.com
Online shopping center for 72-hour Go-Paks.

www.ready.gov/america/getakit/index.html
Suggestions for an emergency supply kit and an instructional video.

www.thebeehive.org/emergencies/be-prepared
Suggested items to include in Go-Paks (including additional items for persons with special needs).

Children and Disasters

www.deep.med.miami.edu/x466.xml
From the DEEP website. Link to the downloadable book, *Children: Stress, Trauma and Disasters*, a resource aimed at explaining the psychological impact of traumatic events on children and families.

www.dhs.gov/xprevprot/programs/gc_1183486267373.shtm
From the DHS website. Information on funding, training, and resources for schools concerning emergency preparedness.

www.fema.gov/kids
Educational tools for parents and teachers to educate children about types of disasters, how to prepare for them, and how to make an emergency kit for yourself and your family.

www.floridadisaster.org/kids/index2.htm
Educational tools and resources for families and schools to educate children about emergencies and disasters. Includes games, downloadable videos and books, and links to other child-oriented websites.

newsanchormom.blogspot.com/2008/12/preparing-children-for-disaster.html
Blog created by an Illinois anchorwoman to bring up-to-date information to mothers on a variety of news topics. Blog for December 12, 2008, covers preparing your children and your home for disasters.

www.sesameworkshop.org/initiatives/emotion/ready
Grover and Zoe learn how to be prepared for emergencies and what to put in an emergency kit (aimed at children ages 3 to 5).

www.slideshare.net/renaissancedoc/rinse-lather-repeat
Preparedness curriculum for elementary school teachers to educate children about emergency planning, Go-Paks, and emergency response.

www.slideshare.net/renaissancedoc/rinse-lather-repeat-and-your-foundation
Outlines the funding requirements for a foundation to back the previous curriculum.

Disasters and Special Needs Populations

Websites about dealing with special needs groups in disasters.

www.chefnoah.com/special_needs.htm
Independent website suggestions for considering special needs in preparing families for disasters and power outages.

www.fema.gov/plan/prepare/specialplans.shtm
From the FEMA website. Suggestions to prepare and care for persons with special needs in disasters, including a booklet to download, reference guide for providers, and links to other government resources.

www.thebeehive.org/emergencies/be-prepared/prepare-go-bag/go-bags-people-disabilities
List of items to add to Go-Paks for persons with special needs.

www.thebeehive.org/sites/default/files/content/site_content/Supplies-and-special-equipment.pdf
Example checklist of items to keep ready for emergencies and to share with your emergency contacts for special needs persons.

Financial Assistance for Individuals/Families

www.disasterhelp.gov/start.shtm
Separate website from main FEMA website. Explains the disaster assistance program, including eligibility, how to apply, and steps in the process. Also provides current information on the status of active disasters.

www.fsa.usda.gov/FSA/webapp?area=home&subject=diap&topic=landing
Official site of the Farm Service Agency (FSA) of the U.S. Department of Agriculture (USDA). Provides detailed information on farm assistance programs in the wake of disaster losses.

www.sba.gov/services/disasterassistance
Small Business Administration can provide low-interest loans for individuals, businesses, and nonprofit organizations for property or assets damaged or destroyed by a declared disaster.

Businesses and Disasters

www.thebci.org
Online resource for a pre-eminent Business Continuity Institute recognized for good practice and professionalism.

www.bomasf.org
Online resource for The Building Owners and Managers Association of California (BOMA California) is a federation of all eight metropolitan BOMA local associations and serves as the collective membership's legislative and regulatory advocate.

www.brma.com
Online resource for Business Recovery Managers Association (BRMA), located in the greater San Francisco Bay Area, devoted to the advancement of the theory and practice of business recovery, disaster recovery, contingency and continuity planning, and emergency response management.

www.cidrapsource.com/source/index.html
Resource for business preparedness and continuity planning in the context of infectious diseases and public health crises (University of Minnesota).

www.contingencyplanning.com
Contingency Planning & Management provides educational and training services to the emergency management, security, and business continuity professionals.

www.disaster-resource.com
Online guide to finding information, organizations, vendors, and resources to assist businesses to maintain continuity in the wake of a disaster.

https://www.drii.org
Website of the Disaster Recovery Institute International, an organization providing education and professional certification for Business Continuity Professionals.

www.drj.com
Disaster Recovery Journal is the most widely read Business Continuity publication in the world.

www.globalcontinuity.com
Business risk assessment and continuity.

www.ibhs.org/business_protection
Toolkits and other products from the Institute for Business & Home Safety to assist small businesses in preparing for disasters.

www.osha.org
Official website of the Occupational Safety and Health Administration, Department of Labor. Searching the A-Z index, you can find these disaster-related listings: Disaster Recovery Fact Sheets, Disaster Site Worker Outreach Training Program, Emergency Preparedness and Response, and Hurricane Assistance and Recovery.

www.sba.gov/services/disasterassistance
Small Business Administration's disaster assistance website featuring aid available only through the SBA, including low-interest loans and grants.

Weather and Disaster Alerts

www.gdacs.org
The Global Disaster Alert and Coordination System provides near real-time alerts about natural disasters around the world and tools to facilitate response coordination, including media monitoring, map catalogs, and a Virtual On-Site Operations Coordination Centre.

www.nhc.noaa.gov/index.shtml
The National Weather Service National Hurricane Center website.

www.stormpulse.com
A compilation website that provides weather information in an interactive fashion with an emphasis on hurricanes.

www.weather.com
Official website of The Weather Channel.

www.wunderground.com
Online resource utilizing the most extensive Internet community of webcams, weather professionals, and private weather stations to provide real-time weather information from around the globe.

www.cpc.ncep.noaa.gov/products/predictions/threats/threats.php
The National Weather Service National Climate Prediction Center website.

www.firedetect.noaa.gov/viewer.htm
The National Oceanographic and Atmospheric Administration Satellite Wildfire detection website.

www.spc.noaa.gov/exper/firecomp
The National Oceanographic and Atmospheric Administration Fire Weather Analysis website.

www.nifc.gov/fire_info/nfn.htm
The National Interagency Fire Center (NIFC) website is the nation's support center for wildland firefighting. Eight different agencies and organizations are part of NIFC.

drought.unl.edu/dm/monitor.html
National Drought Mitigation Center website based at the University of Nebraska-Lincoln.

www.hpc.ncep.noaa.gov/qpf/excess_rain.shtml
The National Weather Service rainfall prediction website.

volcano.wr.usgs.gov/activity/index.php
The United States Geological Survey volcano activity website.

volcanoworld.wordpress.com/category/volcano-activity-reports
Volcano activity reports from around the world in blog report format.

www.ndbc.noaa.gov/rmd.shtml
The National Oceanic and Atmospheric Administration's National Data Buoy Center website.

www.arl.noaa.gov/index.php
The National Oceanic and Atmospheric Administration's Air Quality and Climate website.

Community Resources

www.deep.med.miami.edu/x21.xml
From the DEEP website. Downloadable training manual for community disaster training.

dir.yahoo.com/Society_and_Culture/Environment_and_Nature/ disasters
Online resource for Yahoo! groups and Yahoo! communities.

www.linkedin.com/groups?gid=50102
American Academy of Disaster Medicine group on LinkedIn.

www.linkedin.com/groups?gid=50103
Disaster Healthcare group on LinkedIn.

www.linkedin.com/groups?gid=59201
Disaster Law group on LinkedIn.

www.linkedin.com/groups?gid=64823
Personnel Continuity & Resilience group on LinkedIn.

www.linkedin.com/groups?gid=2990
American Red Cross group on LinkedIn.

www.linkedin.com/groups?gid=95914
Crisis, Emergency & Disaster Recovery Professionals group on LinkedIn.

www.linkedin.com/groups?gid=89832
Emergency Management and Homeland Security Professionals group on LinkedIn.

Government Reports

books.nap.edu/openbook.php?isbn=0309101719
Government report on the state of emergency medical care for children featuring a chapter on disaster preparedness in pediatric hospitals.

books.nap.edu/openbook.php?isbn=0309101743
Government report on the state of emergency medical services (EMS) featuring a chapter on disaster preparedness coordination between hospitals and community rescue services.

books.nap.edu/openbook.php?isbn=0309101735
Government report on the state of emergency medical care for adults featuring a chapter on disaster preparedness in hospitals.

Faith-Based Resources

www.elca.org/Our-Faith-In-Action/Responding-to-the-World/Disaster-Response.aspx
Disaster response website for the Evangelical Lutheran Church of America.

www.mds.mennonite.net
Disaster response website for the Mennonite Disaster Service.

www.pcusa.org/pda
Disaster response website for the Presbyterian Disaster Assistance.

www.nvoad.org
Disaster response website for the National Voluntary Organizations Active in Disaster (National VOAD), a forum where organizations share knowledge and resources throughout the disaster cycle—preparation, response, and recovery—to help disaster survivors and their communities.

www.namb.net/site/c.9qKILUOzEpH/b.224451/k.A400/ Disaster_Relief.htm
Disaster response website for Southern Baptist Disaster Relief.

www.catholiccharitiesusa.org/NetCommunity/Page.aspx?pid=305
Disaster response website for Catholic Charities USA Disaster Response.

www.jdc.org/jcdr_main.html
Disaster response website for the Jewish Coalition for Disaster Relief.

www.providentliving.org
Website of the Church of Jesus Christ of Latter-Day Saints. See the Emergency Preparedness and Response link for suggestions on family emergency plans, food storage, and disaster cleanup safety.

streaming.byui.edu/safetyoffice/flu.wmv
An 18-minute training video on Avian Influenza (Bird Flu), prepared for students and employees of Brigham Young University, Idaho.

Animals and Disasters

www.avma.org/disaster
Site of the American Veterinary Medical Association (AVMA). Wealth of information concerning disaster preparedness and disaster care of animals for veterinarians, emergency planners, and owners alike. Includes downloadable resources and an interactive map to locate state resources.

Recommendations for preparedness, sheltering, evacuation, and post-evacuation needs of pets and livestock (includes downloadable brochures).

www.bt.cdc.gov/disasters/petshelters.asp
From the CDC website. Information on care and shelter of animals during disasters. Includes downloadable booklets, brochures, PowerPoint presentations, and a link to Pets 911 (online lost animal locator resources).

www.disasterhelp.gov/gathering-loved-ones.shtm
Separate website from the main FEMA website. Page provides links to caring for and finding lost animals after disasters.

emc.ornl.gov/CSEPPweb/animals/animals_emergencies_owner.htm
From the Oak Ridge National Laboratory. Video reviews important information on planning for the evacuation, care, and shelter needs of pets and livestock during a disaster (including animal Go-Paks).

www.fema.gov/plan/prepare/plan.shtm
From the FEMA website. Links to suggestions for preparing for the needs of pets and livestock during disasters.

**www.hsus.org/hsus_field/hsus_disaster_center/resources/
disaster_preparedness_for_pets.html**
Site of The Humane Society of the United States (HSUS).

Books

Crisis Preparedness Handbook, by Jack A. Spigarelli

Outlines preparedness, accumulating supplies, getting mentally and physically prepared, and having your finances in order. This book is mostly about food preparation for a major disaster, with emphasis on having a year's food storage, milling your own grain, growing sprouts, home canning, dehydrating, freeze-drying, etc. The book offers advice on other topics, including weapons, hand tools, clothing, energy, medical, sanitation, transportation, communications, and home preparation.

Preparedness Now!, by Aton Edwards

Preparedness book organized into brief chapters on a variety of important topics, including: water, food, shelter, sanitation, communication, transportation, and protection. It is also filled with many packing lists detailing what you should get in preparation. It introduces the e-kit and grab-n-go bag with more extensive items.

Emergency Food Storage & Survival Handbook, by Peggy Layton

This book offers summaries on preparing for short-term emergencies, discusses how to store and purify water, talks about setting up a food-storage program, and has some suggestions for how to store food.

Organize for Disaster, by Judith Kolberg

Emphasis is on understanding the federal resources (i.e., FEMA, Red Cross, etc.) available, creating a personal intelligence network, organizing essential documents, maintaining insurance coverage, listing a home inventory, preparing your house for disaster, basic first aid, and having a good family communication plan.

Making the Best of Basics, Family Preparedness Handbook, by James Talmage Stevens

This book is almost completely about in-home food storage and preparation. There is little discussion outside that (except for water issues).

No Such Thing As Doomsday, by Philip L. Hoag

Well-researched insights into disaster preparedness including water, food, heating/cooking, light, power, communications, medical, sanitation, and security.

Disaster Prep 101, by Paul Purcell

One of the most detailed disaster preparedness books for the lay public. This book is extremely detailed with instructions and diagrams covering virtually every possible contingency that could befall a family during a disaster. The author states that this book is designed to not only help prepare the family for disaster, but to be the one reference the family takes with them "just in case."

The Edge of Disaster: Rebuilding a Resilient Nation, by Stephen Flynn

An eye-opening book detailing the state of vulnerability in the United States and the lack of a culture of preparedness. Well written and researched, the book concentrates on the argument for preparedness.

Disaster Response (Library in a Book), by Fred C. Pampel

One of the Library in a Book series, this is a definitive reference for the dizzying array of jargon, acronyms, programs, agencies, and aide.

Surviving a Disaster, Evacuation Strategies and Emergency Kits for Staying Alive, by Tony Nester

The third in the Practical Survival Series provides a review of disaster planning, scenarios, and the all-important emergency kit (Go-Pak). Includes sections for preparing children and for disaster responders.

Disaster Survival: A Common Sense Guide to Catastrophe Preparedness, by L. R. Moehsmer

This book guides the reader through the steps needed to become self-sufficient for a short-term disaster situation. It also encourages the reader to add their own ideas to customize their plan of action.

Don't Get Caught with Your Pantry Down!, by James T. Stevens

This 1998 guide is designed to help consumers find the required preparedness resources—information, products, emergency supplies, long-term storage foods, foodstuffs, items of equipment, and services—all items necessary to achieve any level of personal, family, or organizational preparedness. This pre-9/11 guide is as fresh today as when it was written.

Emergency Food in a Nutshell, by Leslie Probert and Lisa Harkness

This book keeps gathering and using food storage SIMPLE and DO-ABLE! Simplified shelf-life and water storage information, from foremost U.S. experts, gives you the bottom line on these issues.

Appendix B

Go-Pak Checklists

This appendix contains the same information about what to pack in your Go-Pak as appears in Chapter 3, but in an easy-to-use checklist format.

Critical Information and Important Papers

- ❏ Driver's license
- ❏ Insurance policy documents
- ❏ Medical insurance cards and information
- ❏ Vehicle and real estate titles
- ❏ Birth certificates, Social Security cards, and passports
- ❏ Credit cards
- ❏ School records for everyone in the family who's in school
- ❏ Medical and immunization records for everyone
- ❏ Mortgage boat, auto, and personal loan information, including ID number, monthly payment amount, company contact information, and address
- ❏ Credit card numbers, including exact name on the card, expiration date, authorization code, company contact numbers, and monthly due date

- ❏ Bank and brokerage account numbers and the branch names and contact numbers
- ❏ Marriage certificate and custody documents

Personal Clothing

- ❏ Hat (to prevent sunburn and provide shade, and to keep the rain out of your face)
- ❏ Closed-toe shoes (to protect your toes)
- ❏ Jacket with hood
- ❏ Poncho (or raincoat and pants)
- ❏ Long trousers, one pair (to protect your legs)
- ❏ Shorts, one pair (cooler for hot weather)
- ❏ Long-sleeved shirt, one (to protect arms and prevent sunburn)
- ❏ Short-sleeved shirts, three (cooler for hot weather)
- ❏ Underwear, three or more
- ❏ Socks, three pairs or more
- ❏ Large bandanna, one or more (to prevent sunburn)

Miscellaneous Items

- ❏ Lock and key (to keep the Go-Pak locked when you're not using it)
- ❏ Personal first aid kit
- ❏ Extra sets of car and house keys
- ❏ Knife, lockable
- ❏ Multi-use tool (such as a Leatherman or Gerber)
- ❏ Work gloves (to protect your hands during cleanup or heavy work)
- ❏ Face mask or dust mask (to protect your lungs during cleanup or heavy work)
- ❏ Flashlight and batteries
- ❏ Whistle (to signal and call for help)
- ❏ Compass

Personal Gear

- ❏ Waterproof watch, preferably with an alarm
- ❏ Sunglasses
- ❏ Extra set of prescription glasses
- ❏ Personal hygiene items
- ❏ Bar soap
- ❏ Shampoo
- ❏ Razor blades
- ❏ Shaving cream
- ❏ Deodorant, unscented
- ❏ Toothbrush, toothpaste, dental floss
- ❏ Comb, brush, and unbreakable mirror
- ❏ Toilet paper
- ❏ Lip balm
- ❏ Hand lotion, unscented
- ❏ Sunscreen, unscented
- ❏ Insect repellent, pump type, unscented
- ❏ Flip-flops or some other shower shoe
- ❏ Biodegradable laundry detergent
- ❏ Anti-fungal foot powder and moleskin (for blisters)

Medications

- ❏ Prescriptions (a two-week supply at least)
- ❏ Aspirin
- ❏ Antacids (Maalox/Mylanta/TUMS)
- ❏ Antihistamines (Benadryl or Zyrtec)
- ❏ Tylenol
- ❏ Decongestants
- ❏ Imodium (anti-diarrheal medication)

Miscellaneous Items

- ❏ Towels (large and small)
- ❏ Washcloth
- ❏ Books, reading material, crosswords, Sudoku
- ❏ Inflatable pillow
- ❏ Handi-wipes, unscented
- ❏ Laundry bag and clothespins
- ❏ Safety pins and sewing kit
- ❏ Waterproof matches and fire starter material
- ❏ Electrical tape
- ❏ Parachute rope, 20 feet
- ❏ Extra shoelaces
- ❏ Water purification tablets
- ❏ About 72 hours of high-nutrition snacks and pocket snacks (such as trail mix, gum, dried fruit, and hard candy)
- ❏ Notebook, pens, pencils
- ❏ Cup (pocket size, collapsible)

Communication Supplies

- ❏ Small portable radio with headphones
- ❏ Roll of quarters for a payphone
- ❏ Cell phone charger (12-volt car and 110-volt wall outlet)

Additional Considerations for Children

- ❏ Photos of family members with names written on the back
- ❏ Names and contact information for at least four family members and friends
- ❏ A list of local information sources, such as TV and radio stations, weather services, local access cable, local government emergency operation center (EOC) numbers, and local print media

Writing Your Family Disaster Plan

This appendix is a guide to writing a disaster plan specifically tailored for your family's needs. A family disaster plan ensures that you know who's doing what in an emergency, sets up lines of communication, and gives you a tool that helps you remember to do the things you need to do.

What Your Disaster Plan Should Contain

At a minimum, your disaster plan should contain the following information:

- Participant contact information

- Safety buddies

- Meeting places

- Emergency services numbers

- Family contact points

- Medical and special needs information

- Important documents

- Emergency agencies

- Home and business contacts

You can download sample forms for creating a family disaster plan through www.mauricearamirez.com or www.hedtke.com/disaster.htm.

Participant Contact Information

The first thing in your family disaster plan is a list of names, addresses, phone numbers, and e-mail for each participant. As you've seen in this book, "family" doesn't just mean your immediate blood relatives, but can include friends, neighbors, and co-workers. List everyone who is participating in this plan: you, the other people in your house, and any friends, neighbors, or co-workers who are going to be part of your tribe in a disaster. It's a good idea to list relationships and ages, too. Include photos of everyone. Even if you're printing scanned copies of black-and-white photos, they're better for tracking people than trying to describe them to someone by their age, height, and hair color.

Safety Buddies

Pairing people up with a "safety buddy" doesn't just apply to kids (who certainly need a lot of adult supervision in a disaster to keep them safe). Adults should also be paired up with someone so that nobody has to do anything by themselves. Assigning safety buddies and doing things together probably saves more lives in a disaster than any other disaster planning technique. List the assigned safety buddy next to each participant's name. If the safety buddy changes, write the new name into the disaster plan right then and there.

Meeting Places

Although your family may be all together when a disaster strikes, more often than not, one or more family members will be at work, school, or elsewhere. It is vital to have local and out-of-town meeting places for everyone to travel to when separated.

The disaster plan should list at least one meeting place in town and out of town, as well as at least one local and out-of-town shelter. If you have multiple meeting places, assign a priority to them: "If possible, go here first. If this site is unreachable or unsafe, go there." Where possible, you should also have a system for leaving a message at one meeting place to show when people were there and where they've gone.

Emergency Services Numbers

Provide a list of standard emergency numbers, including police and fire emergency and nonemergency numbers, the local Office of Emergency Management, the local Red Cross branch, and ambulance services and hospitals.

Family Contact Point

Provide complete contact information for the persons designated as the contact point for messages and information. If possible, have in-town and out-of-town contact points. You may want to have more than one out-of-town contact point so that if one contact point is unreachable, the secondary contact point can fill in. Contact points should be reasonably available and be able to relay messages and act as emergency contacts for everyone participating in the disaster plan. If you get separated from the rest of the group, you should call in to the family contact point until you're reunited with the rest of the family.

Medical and Special Needs Information

Everyone in the family needs to have their medical information in the disaster plan. Children, the elderly, and people with medical needs particularly require continuity of health care, as well as having all the special "stuff" to keep them healthy and comfortable.

List contact information for doctors, specialists, clinics or hospitals, home health-care workers, mental health professionals, dentists, and local and mail-order pharmacies. If you have pets, list the veterinarians.

For each participant, list any medications they're on, the dose, and the directions. Also list any allergies and the reactions.

For special needs patients, list such things as the medic alert ID, any special requirements such as food, assistive devices, and treatment regimens. In addition, list the models, serial numbers, and suppliers for medical equipment such as pacemakers, feeding pumps, home IV units, suction machines, wheelchairs, and Braille or lower vision equipment.

Important Documents

As you saw in Chapter 3, your Go-Pak will already have a number of important documents. However, you may want to include photocopies in the disaster plan of some of the most essential, such as your driver's license or other personal identification, insurance documents, medical insurance cards, vehicle and real estate titles, birth certificates or adoption papers, name change documents, Social Security cards, and passports. Having photocopies of information in the disaster plan is convenient, but it may also be a risk if the disaster plan is lost or stolen.

Emergency Agencies

You should list contact information for some standard emergency agencies, such as FEMA, the American Red Cross, and your local disaster assistance center. In addition, it's likely that your city, county, or state may have disaster relief agencies as well as agencies dealing with specific disasters, such as wildfires, earthquakes, or floods. Check the resource lists in Appendix A and gather contact information for disaster relief and assistance agencies in your area.

Home and Business Contacts

List contact information for people and companies you'll need for getting your home repaired and your business back on its feet, such as insurance agents and policy numbers, plumbers, electricians, roofers, and general contractors. Also include contact information for landlords or property management firms and corporate HQ if your business is part of a larger company.

Distributing the Disaster Plan

Every participant gets their own copy of the disaster plan. A good method is to print multiple copies and insert the pages into plastic sheet protectors (so they don't get torn out) in a plain, three-ring binder. Put the person's name prominently on the binder of each disaster plan. Store each disaster plan with each person's Go-Pak so that it's instantly available in case of emergency.

You can get fancier if you like: for example, you can also include all the information on an inexpensive thumb drive in a plastic container. Many varieties are made to be worn on wristbands or as necklaces and are reasonably waterproof. The thumb drive can also hold more family photos as well as additional personal, financial, and medical records.

Evacuation/Sheltering in Place Checklists

The checklists here should give you enough basic information to work with, but you should modify these to suit your own situation. For example, if you have a family member with special needs, you'll need to add tasks relating to medical or transportation requirements. You should have a checklist for each phase of your evacuation or sheltering in place. Make sure everyone has a copy of the checklist information with them.

Preparing to Evacuate or Shelter in Place

This checklist covers the tasks for preparing to evacuate or to shelter in place.

People

- ❑ Everyone knows the destination and how to get there.
- ❑ Everyone knows their role in the plan.
- ❑ Everyone has a printed copy and a thumb drive of maps, phone numbers, photographs, and contact schedules and information.

❑ Everyone has their Go-Pak with them.

❑ Everyone knows how to reach the contact point.

Pet

❑ Take pet's Go-Pak.

❑ Crate pet and get ready for a shelter, hotel, or kennel.

❑ Take vaccination records and animal tags.

Stuff

❑ Secure valuables in safe or safe-deposit box.

Home

❑ Power turned off.

❑ Water turned off.

❑ Gas turned off.

❑ Unplug all electric appliances.

❑ Disconnect phones from wall.

❑ Disconnect cable TV/satellite converter boxes and cable modem from cable.

❑ Disconnect computers from Internet connections.

❑ Fill clean water jugs.

❑ Fill swimming pool.

❑ Place all outdoor furniture and appliances indoors.

❑ Cover/wrap and secure all outdoor items too large to bring indoors.

❑ Fill bathtubs for extra water.

❑ Secure storm shutters and/or window planks.

❑ Place trashcans indoors.

❑ Tie down or chain all gates and fences.

Boat

- ❏ Stow all loose gear.
- ❏ Add old tires and bumpers around the boat to prevent impact damage.
- ❏ Add chafing sleeves to all lines.
- ❏ Tie up extra lines.
- ❏ Lash down, bolt down, and secure all loose equipment.
- ❏ Charge batteries on automatic bilge pump.
- ❏ Seal all openings with duct tape.

Car

- ❏ Fill gas tank.
- ❏ Check tire pressure, *including the spare.*
- ❏ Check all fluid levels (oil, washer fluid, antifreeze/coolant).

Returning to Normal Checklist

This checklist covers the tasks for returning to normal from evacuation or sheltering in place.

People

- ❏ Everyone knows the potential dangers at home.
- ❏ Everyone knows their role upon returning home.
- ❏ Everyone has a printed copy and a thumb drive of maps, phone numbers, photographs, and contact schedules and information.
- ❏ Everyone has their Go-Pak with them.
- ❏ Everyone knows how to reach the contact point.

Pet

- ❏ Take pet's Go-Pak.
- ❏ Crate pet until safe and secure at home.
- ❏ Take vaccination records and animal tags.

Home

- ❏ Turn power on.
- ❏ Turn water on.
- ❏ Turn gas on and *immediately* light pilots.
- ❏ Plug-in appliances and connect to cable/phone line/Internet.
- ❏ Connect computers to Internet connections.
- ❏ Connect cable TV/satellite converter boxes and cable modem to cable.
- ❏ Connect phones to wall.
- ❏ Plug in all electric appliances.
- ❏ Refill and chlorinate swimming pool.
- ❏ Move outdoor furniture and appliances outdoors.
- ❏ Uncover/unwrap all other outdoor items.
- ❏ Don't drain bathtubs or empty water jugs until water supply is declared safe.
- ❏ Remove storm shutters and/or window planks.
- ❏ Place trashcans in their usual location.
- ❏ Untie/unchain all gates and fences.

Stuff

- ❏ Check on valuables in safe or safe deposit box.

Boat

- ❏ Remove old tires and bumpers from the boat.
- ❏ Unseal all openings sealed for the storm.

Car

- ❏ Refill gas tank.
- ❏ Check tire pressure, *including the spare.*
- ❏ Check all fluid levels (oil, washer fluid, antifreeze/coolant).

Index

166